Solutions for Soil and Structural Systems using Excel and VBA Programs

Solutions for Soil and Structural Systems using Excel and VBA Programs

Robert L. Sogge

PE, PhD, Civil Engineering Consultant, Tucson, USA

A John Wiley & Sons, Ltd., Publication

This edition first published 2012
© 2012 John Wiley & Sons Ltd

Registered office
John Wiley & Sons Ltd, The Atrium, Southern Gate, Chichester, West Sussex, PO19 8SQ, United Kingdom

For details of our global editorial offices, for customer services and for information about how to apply for permission to reuse the copyright material in this book please see our website at www.wiley.com.

Library of Congress Cataloging-in-Publication Data

Sogge, Robert L.
  Solutions for soil and structural systems using Excel and VBA programs / Robert L. Sogge.
      p. cm.
  Includes bibliographical references and index.
  ISBN 978-1-119-95155-1 (hardback)
  1. Soil mechanics--Data processing.   2. Finite element method--Computer programs.   3. Microsoft
Visual BASIC.   4. Microsoft Excel (Computer file)   I. Title.
  TA710.S586 2012
  624.1'51360285554--dc23

                                                                                    2012004274

A catalogue record for this book is available from the British Library.

Print ISBN: 9781119951551

Set in 10/12.5pt Palatino by Laserwords Private Limited, Chennai, India
Printed in Malaysia by Ho Printing (M) Sdn Bhd

To Ralph M. Richard and Hassan A. Sultan, Professors at the University of Arizona, who shared their ideas and approaches and turned me on to how they thought. With people who take their job as seriously as they do, we can have an educated future generation. Their skill in the presentation and innovation shown in addressing new topics stood out in the university environment.

Also in memory of Christian Sogge, who always asked me, "What do you want to be when you grow up – a doctor or a lawyer?" and I always answered, "I want to do what you do."

Contents

About the Author

Robert Sogge is a registered Professional Engineer, PE Civil, in the states of Arizona and California. He has a PhD in Civil Engineering, specializing in soils and structures, from the University of Arizona. His training at that school, Northwestern University (BS), and the University of New Mexico (MS) was from some of the finest professors in the land, including Osterberg, Krizek, Kondner, Yao, Triandafilidis, Sultan, and Richard. Such inspirational instruction led first to an academic career in teaching and research. He has published papers in the specific area of soil–structure interaction and the use of computer methods to implement the analysis of anchored sheet and laterally loaded piles.

Dr. Sogge wrote his first programs in FORTRAN for mainframes and was part of the evolution of computer languages that occurred with the development of personal computers, specifically the Osborne 1 microcomputer in 1981, and culminating with spreadsheet software that integrated the VBA language into the Excel spreadsheet program. This research was then applied to practical consulting for projects on raft-slab foundation design, drilled pier foundation systems for bridges in a water environment, shoring designs for deep excavations, and buried cast-in-place reinforced concrete arch culvert structures, all employing computer stress analysis procedures. Many of the programs included in this text were developed throughout his career on these consulting jobs.

Dr. Sogge has owned a firm that provided design-build services for buried arch culverts, and has lectured to various state highway departments and at educational seminars conducted by the International Association of Foundation Drilling. He is a local and national member of both the American Society of Civil Engineers and the Sigma Xi scientific research society. He lives in Tucson, Arizona where he runs a specialized consulting engineering business providing expertise on the topics noted above.

Preface

(explanation to users on how they can best use text and what they can expect to learn from it)

This book is intended as a textbook for students and a reference book for practicing professional geotechnical and structural engineers. Its contribution is in developing those topics in structural analysis and those required from soil mechanics to present an analysis of systems involving soil–structure interactions.

An ancillary purpose is to present the basics of Excel spreadsheets and programming of; Visual Basic for Applications (VBA) macros by demonstrating various techniques in the analytical spreadsheets and programs provided.

The units used in this text are US Customary Standard (fps) units.

Simple beam–bar models are used with the PFrame finite element program to analyze a variety of examples in the soil–structure interaction arena.

Robert L. Sogge
Tucson, Arizona
December 2011

Acknowledgments

As well as the two professors, my appreciation goes out to the following people:

To all the clients, too numerous to mention, whose projects have provided the foundation on which this book is built. Included among them is Conrad Huss and Dan Neff of M3 Engineering, Wolf Michelson of Hydro Arch and Mike Cable of Integra Arch.

Dan Stevens, former classmate at Northwestern University, a life-long learner who provided long-distance help on initial aspects of this book.

Helen Ireland, who showed me that a dream I had, using recent computer analysis approaches, could become a reality.

Jason Morrish, a consultant extraordinaire, who provided easier ways to start programs, input values, and plot lines.

Zbig Osmolski, who interrupted his and my vacation in Rio to offer some helpful suggestions.

Christin Sogge and Nick Garwood, who both were always ready for an Excel discussion, posed some insightful questions that helped me present some ideas, and made some useful comments.

Art Gould, who gave guidance, encouragement, and support to go for publication and how to do it, Mike Kovara, who gave an engineer's reading to the text in its final proof stages.

And, of course, probably the most important, the Wiley staff – specifically Liz Wingett, Project Editor for Engineering books, Chichester (lwingett@wiley.com) who saw this book through all the steps required to get it to publication; Paul Petralia, Senior Editor (USA), Sophia Travis – Chichester; Sandra Grayson – Chichester; Clarissa Lim, Senior Production Editor, John Wiley & Sons, Singapore (cllim@wiley.com); Neville Hankins, editor, Bristol, UK; Kalpana

Madhavan, typesetter, Chennai, India. They did it – they put it together and made it happen.

Thank you all.

R.L. Sogge
Tucson, Arizona
December 2011

Part One
Computer Software

"A hundred times every day I remind myself that my inner and outer life, depend on the labours of other men, living and dead, and that I must exert myself in order to give in the same measure as I have received."

– Albert Einstein

1 Microsoft Excel Spreadsheet
2 Microsoft VBA Programming Language

1

Microsoft Excel Spreadsheet

The two computational essentials presented in this text are the Excel spreadsheet and the programmed procedures that, when written in the VBA (Visual Basic for Applications) macro programming language and attached to a workbook, can handle any programming chore. This text will concentrate on the engineering applications of structures and soils and their interaction. Application to any other area of specialization will be evident. The text will approach the teaching not inductively, with instructions in general theory, but deductively, with examples showing the specific aspect being addressed. The book is a learning tool as much as it is a reference guide, so it should be used in conjunction with the enclosed programs using the data in the examples, or other data created by the user to solve problems on the computer. To begin, the Excel spreadsheet will be covered in this chapter and the VBA programming language in Chapter 2.

1.1 History of Spreadsheet Development

Excel is a spreadsheet program developed by Microsoft and is part of Microsoft's Office suite. The earliest spreadsheets were VisiCalc and SuperCalc, which were first distributed in 1979 and 1980. Microsoft marketed its Multiplan in 1982 on CP/M-based systems. Shortly after, Lotus 1-2-3 became popular, eventually dominating on the newer MS-DOS systems. Borland's spreadsheet program, Quattro Pro, was also a big seller.

The Microsoft Windows operating system was introduced in 1985. Microsoft's first Windows-based spreadsheet program was introduced in 1987, renamed Excel, and called version 2.0. Eventually, Excel supplanted Lotus 1-2-3 as the industry standard for spreadsheet programs, and continues as such today. The 1993 version of Excel was the first to include VBA. Excel 95 was the first to support a 32-bit operating system. Excel 2007 introduced the Ribbon, a complete revamp of the user interface (UI) which proved to be a contentious change at first.

Other office suites that still feature a spreadsheet continue to be offered: Open Office.org (OOo – formerly StarOffice developed by Sun Microsystems), IBM's Lotus Symphony (which since 2010 has incorporated code from OpenOffice), and

Solutions for Soil and Structural Systems using Excel and VBA Programs, First Edition. Robert L. Sogge.
© 2012 John Wiley & Sons, Ltd. Published 2012 by John Wiley & Sons, Ltd.

LibreOffice, along with Google Docs, the online office suite. These spreadsheets are not a true substitute for the Excel spreadsheet in Microsoft Office, as support for VBA is not found in them, and neither is full compatibility with Microsoft's .doc(x) and .xls(x) formats. That said, many programs are capable of both reading and writing .xls and .xlsx files. The strength and reasonable price of Microsoft Excel software make it the only choice.

1.2 Excel 2010

Beginning with Windows XP, Microsoft provided the choice of either a 32- or 64-bit version for the operating system. It was not until Microsoft Office 2010 that both 32- and 64-bit versions of Excel were offered. The 32-bit version enables the use of solutions built in previous versions of Microsoft Office without modification. CPUs sold today from AMD and Intel are 64/32-bit capable and will run either 32- or 64-bit Windows.

The 64-bit version of Microsoft Office 2010 requires use of the 64-bit Windows 7 operating system on a computer with a 64-bit capable processor. The 64-bit version can address a much larger memory than the 4 GB that a 32-bit processor can. By default the 32-bit version of Microsoft Office 2010 is installed unless a custom installation is performed. Such a version is supported by Microsoft and can handle files created by older versions as well as work on 64-bit operating systems such as Windows 7.

The word spreadsheet is the more common description for what Microsoft Excel refers to as a worksheet. Excel uses the word workbook to refer to a collection of worksheets. If only one worksheet is in a workbook it could be denoted as a worksheet or a workbook. All the worksheets and associated VBA macro procedures in this text have been developed in Excel 2010 but will run in Excel 2003 to 2010. Office 2003 does not read or write the default Office formats of the 2007 or 2010 versions, which is .xlsx for Excel, but the latter versions can readily read files created in the 1997–2003 versions and save them in the old format. Additionally, Microsoft has a Compatibility Pack for Office 2003, available from its Download Center, which allows the older Office suite to read and write the newer formats. The features in Excel 2007 and Excel 2010 are not very different other than the change to offering both 32- and 64-bit versions.

1.2.1 File Conversion and Compatibility

One of the greatest advantages of the Excel–VBA package offered by Microsoft is its backwards and forwards compatibility with previous and new versions. As a transition from a 32-bit Windows XP to a 64-bit Windows 7 operating system is made, the ability to run programs created with the Excel–VBA package still remains. In this transition many users of programs created in one environment find that they cannot be run in the other and are forced to reuse their older computer because the drivers may have issues with the new program.

As stated previously, the reading and writing of Excel 2007 or later file formats using Excel 2003 is possible assuming that the Microsoft Compatibility Pack has been downloaded and installed. The format for an Excel 2007 workbook file is *.xlsx and *.xlsm for a macro-enabled workbook. Office 2007 can open Office 2003 files and then convert and upgrade the *.xls file to either a *.xlsx or *.xlsm file format. Changing from Excel 2003 to Excel 2007 (Office 2007) does not require any change or upgrade to the operating system, such as from Windows XP to Windows 7.

Excel 2010 provides a compatibility check under the File > Check for Issues button. One difference between the 2007 and 2010 versions of Excel is in defining the names for cells. For example, previously Rad2 was an acceptable name, but with the 2010 version it is not. This feature will be covered in more detail in Section 1.6.4 on naming cells.

1.3 Transmitting Cell Values Not Formulas

When sending a spreadsheet to others it may be beneficial to keep the calculations, and specifically the formulas in the cells, private, transmitting only the values in the spreadsheet cells. To copy values in cells that do not contain drawing objects or formulas from one worksheet to another using Excel 2003 requires the steps of selecting the cells and using Paste Special to paste the cell values. Alternatively, Adobe Acrobat, but not the free Adobe Reader, can be used to convert Excel 2003 spreadsheets to PDF files. Such a file allows one to send read-only versions of documents.

In Excel 2007, Microsoft created its own competing version of PDF denoted as XPS. An XPS reader, necessary to open such a file, comes with Windows Vista and Excel 2007, and can be downloaded for free from Microsoft.

Excel 2010 makes copying the values in a spreadsheet simple. Any portion of a worksheet that is set as a Print Area can be saved directly to a PDF file. Alternatively, the values of any portion of a worksheet can be copied using the Copy as Picture feature adjacent to the menu copy button in Excel 2010. By copying in this manner, cell formulas and structures are not copied.

1.4 Accuracy

Excel stores and calculates with 15 significant digits of precision and uses scientific notation for numbers ranging from $E-307$ to $E+307$. This degree of accuracy is the same as variables denoted with Double Precision in the VBA language macros, to be introduced in Chapter 2. Variables denoted as Single Precision in VBA have only seven digits of accuracy. A comparison of this 15-digit accuracy from the same calculation in an Excel worksheet and a VBA macro is presented in the workbook **eCalc**. All workbooks and their worksheets are contained in the DVD that is a part of this text.

Rounding numbers for currency calculations in a spreadsheet to a specific number of decimal places uses the ROUND() function. To round a value to two decimal places, the following statement is used:

```
=ROUND(H18*IntYr/PpY,2)
```

An example of this rounding is shown in the worksheet **LoanPmtSchd** of **LoanProg** in Chapter 24.

1.5 Saving

The best way to save spreadsheets and attached VBA macro programs that are changed or updated is to use the same filename for all those performing a similar function and append with a revision number (1), (2) or -r1, -r2, and so on. The last one or two revisions should be saved in order to return to them, if needed.

When saving an older version file in the new format of Excel 2010, the ranges for the data selected for charting are not transferred and the data ranges must be reselected.

1.6 Implementation of Excel Features

1.6.1 General Tips

Using columns or rows to provide a type of format spacing is not good practice. Formatting of individual cells should be kept to a minimum.

It is best to use only one page orientation, portrait or landscape, between the worksheets of a workbook. Try to keep a consistent print size, scale, and page orientation so information can readily be transferred between cell locations in different worksheets.

Rather than making extensive use of headers and footers it is better to put such information directly in the spreadsheet.

When an analysis style has been developed, stick with it and use it as a template for other analytical spreadsheets. Much of the information does not change between systems being analyzed.

Hiding rows or columns in a worksheet results in less clarity and more confusion.

If the value in cell E24 were set equal to 2 * the value in cell E23, and then cell E23 was replaced by cutting and pasting information from another cell, then cell E24 would become #REF!. The same result would not occur if the value in cell E23 were replaced by copying (and not cutting) and pasting information from another cell.

Absolute cell references, that are permanent if copied or cut, are differentiated by using the $ symbol in front of either the row or column reference, or both. Relative cell references that change when moved do not have these constant row or column references.

When copying a table in Excel to a Microsoft Word 2010 text document, and then converting the Table to Text, either left or right click on the table and almost magically a Table Tools with Design and Layout sub-tabs appears under the Home tab (even if your eye was never up there looking for it). Under the Layout tab is a Convert to Text tab that will allow you to convert the table.

Links to other workbooks can be viewed and deleted by using Edit > Links > Source > Break. It is not advisable to link workbooks.

A series of values can be filled in by dragging the handle at the right of the cell using the right mouse button.

1.6.2 Fonts

Changing the font of a worksheet through Format > Cells > Font rather than through Format > Style > Modify Font yields two different-sized worksheets, even if the same font is used on each.

The Sans Serif Arial font is used for cells since it is more compact both horizontally and vertically than other similar sans-serif fonts like Calibri (a standard in Excel 2010) or Gill Sans MT.

The column width of cells in Excel is governed by the Standard Format font width in the Tools > Style setup page for the spreadsheet.

1.6.3 =IF Statements

In Excel there is no Go To branching, For statements or any kind of sequential steps – instead the entire spreadsheet is computed at once. The =IF statement is used to replace these logical statements that are not present. In Chapter 9, program **Class** has two versions. One version just uses a spreadsheet and the other version employs a VBA macro. The same types of IF statements are used in both in determining the result:

```
=IF(condition, action if True, action if False)
```

Logical functions can be made conditional by the statements AND and OR:

```
=AND (test1, test2) for testing to see if ALL of the
 conditions are True.
=OR (test1, test2) for testing to see if ANY of the
 conditions are True.
```

An example of both IF and AND is in the statement

```
=IF(AND(H721>F721,H721>F720),"OK","Not OK")
```

The statement

```
=IF (B16=1, E16, IF(B16=4,E16,IF(B16=5, E16, 0)))
```

may be written as

```
=IF (OR (B16=1, B16=4, B16=5), E16,0)
```

The statement

```
=IF(AND(B118=1,C118=1),DenA-0.0624, IF(AND(B118=1, C118=0), DenA,
IF(AND(B118=2,C118=1),DenB-0.0624, =IF(AND(B118=2, C118=0)DenB,0))))
```

can alternatively be written as a nested IF statement:

```
=IF(B118=1.AND. C118=1, DenA-0.0624,IF(B118=1.AND.C118,DenA,
 IF(B118=2,.AND.C118=1, DenB-0.0624, IF(B118=2.AND.C118=0, DenB, 0))))
=IF(C89=1,0,IF(D89=1,0,-6*E*I))
```

Examples of nested IF statements are as follows:

```
=IF(N67<0.3,"rigid",IF(N67>1.3,"flexible","intermediate"))
```

```
=IF(C89=1.0,IF(D89=1,0,-6*e*i))
```

Rather than nest several IF statements, it is often easier and clearer to separate the nested operations into a series of worksheet cells with a specific single operation being performed in each cell.

1.6.4 Naming Cells

It is often useful to "name" a cell so that a name rather than a cell location can be referred to in a formula. Doing so makes formulas and equations much easier to understand. One of the differences between the 2007 and 2010 versions of Excel occurs in defining the names for cells. Previously Rad2 and ks1 were acceptable names. With the 2010 version they are not. Now it is necessary to change the variable names Rad2 and ks1 to Radi2 and k. The names Rad2, R2, RR2, Ra1, ks1, and k1 are all unacceptable. IRR cannot be used as a variable since it is the Internal Rate of Return function. Now all numbered names must have at least four characters. In lieu of names containing numbers, an alternative naming convention would be to change the numbers to the letters a, b, and so on. The simplest way to make this change is to use the Name Manager and edit the name. Alternatively, in older versions of Excel, all Rad2 can be found and replaced with Radi2 and then the Rad2 name definition deleted, and Radi2 defined as a name.

1.6.5 Functions

Functions require the use of some simple VBA programming statements so their introduction in this chapter is cursory. Functions that are implemented by Procedures that store VBA code consist of the following two types:

- **Sub Procedures:** These will be covered in Chapter 2. Such procedures do not return a value to a cell unless the code specifically contains such a step.
- **Function Procedures:** Such procedures written in VBA return a value to the cell of a worksheet through the name of the function. Worksheet variable values are transferred to the VBA procedure through variable names enclosed in parentheses. These procedures are not identified as a macro when listing macros but when saving they must be in a macro-enabled workbook.

```
Function C(A as Single, B as Single), as Single
C = SQRT(A + B)
End Function
```

In a worksheet enter the following in a cell

```
= Function C(3, 4)
```

to return a value for C (the value is 5), to that cell. This is different from a Sub Procedure that does not return value. VBA code is entered into a module where they are stored. A good example of a Function Procedure is in the **LoanProg** program presented in Chapter 24. This Function Procedure employs an iterative routine to calculate interest. There will be more discussion of this topic in Chapter 2. In that chapter VBA macro programming will be integrated with Excel worksheets to show how they function as the palette for data input and output as well as program control.

1.6.5.1 Predefined Functions

Excel also provides functions that are predefined formulas or equations. They may be in Math and Trig or Financial or various other fields. The functions contained in the Analysis Toolpak and the Analysis Toolpak –VBA are automatically enabled in Excel versions beginning with 2007.

Matrix algebra operations such as transpose, multiply, and inversion cover arrays defined by a range of rows and columns. When using the math functions MMULT or MINVERSE from the Insert > Function > Math and Trig menu bar in Excel for multiplication or inversion of matrices, do not enter braces { }. First highlight the destination matrix and then enter the =MMULT or =MINVERSE function with the appropriate arrays. This function is entered by pressing Ctrl+Shift+Enter simultaneously. The braces will automatically be entered during the CtrlShiftEnter step. They indicate an array formula.

Examples that show the use of the MINVERSE and MMULT functions available in Excel are Examples 6.1a, 6.2a, 7.1a, 16.2b and 18.7b for a laterally loaded pile. These examples show two different ways of solving a problem: the usual way using program **PFrame** and using the MINVERSE and MMULT functions available in Excel. This alternative workbook approach sets up the constitutive equations and solves them directly using the Excel functions.

1.6.6 Drawing

In creating a drawing in Excel, always use the drawing menu items rather than relying on Format > Cells > Border to create a "line" image since cell spacing can change and border information may not then properly fit. To simultaneously delete many drawing objects such as lines and arrows, in Excel 2010 use Home > Find & Select > Select Objects > a rectangle to enclose and select all objects being deleted and then a delete key sequence.

It seems there is a slight compatibility problem that exists between the 2003 and 2007 or 2010 versions of Excel with worksheets that contain drawings. Various users have noted that the drawings can be stripped when a file contains them. This usually occurs when drawings are created in two different versions. This problem was not apparent in Excel 2003 worksheets.

1.6.7 Charting

The chart type used exclusively in this book is XY (Scatter). The Chart feature of Excel does not fit all situations. It cannot be readily used to graph or plot a series of lines such as when trying to plot a structure's configuration. For each line segment the endpoints must be given and one ends up with as many data sets defined as there are number of line segments. Therefore it is impossible to set up the plot generally for any structure comprising a variable number of line segments. In Chapter 2 a method will be presented whereby such lines can be readily plotted to the Excel worksheet using a VBA macro program.

The fitting of a curve through the plotted data can be performed using the Add Trendline feature of Excel Charts. Right click on the curve and the option of an Add Trendline appears with the following choices:

Linear $y = ax + b$
Logarithmic $y = a \ln x + b$
Polynomial $y = ax^n + b$
Power $y = ax^b$
Exponential $y = a\, e^{bx}$
Moving Average no equation, an n period $= 2 - 19$.

The equation can be plotted on the chart using the Options box under Add Trendline.

Keyboard Shortcuts
F1 – Help
F2 – Activates the information in a cell for Editing
F3 – Paste name from workbook
F5 – Go To
F7 – Spelling and Grammar
F11 – New chart worksheet
F12 – Save As

File Operation Shortcuts – initiated by Ctrl key plus letter (simultaneously):
Ctrl N – New
Ctrl O – Open
Ctrl P – Print
Ctrl S – Save

Cell Editing Shortcuts
Ctrl ~ – Displays formulas in every cell
Ctrl 1 – Format cell
Ctrl C – Copy
Ctrl F – Find
Ctrl G – Go To
Ctrl H – Replace
Ctrl V – Paste
Ctrl X – Cut
Ctrl Y – Redo last action
Ctrl Z – Undo last action

Related Workbook on DVD

eCalc with VBA macro stored in Module1.

Further Readings

Bloch, S. (2000) *Excel for Engineers and Scientists*, John Wiley & Sons, Inc.
Gottfried, B.S. (2003) *Spreadsheet Tools for Engineers Using Excel*, McGraw-Hill.

2

Microsoft VBA Programming Language

2.1 History of the BASIC Computer Language

Any discussion of VBA begins with the BASIC computer language from which it was derived. BASIC was developed as a simple programming language in 1964 from the then present traditional languages, FORTRAN and ALGOL. At that time most civil engineering analysis programs were written in FORTRAN and run on mainframe computers.

Development of the BASIC computer language for personal computers can be broken into the following three stages:

BASIC with CP/M and DOS
Visual BASIC (Windows)
VBA and Excel with Windows

2.1.1 Stage I – BASIC with CP/M and DOS

BASIC is an interpretive language. It takes the program developed by a programmer and then pre-compiles it into pseudo-code that can be indirectly executed by the interpreter. This can be contrasted with a compiled language in which a program is converted into "machine code" and then directly executed by the CPU.

Apple computers came with Applesoft BASIC, a product licensed from Microsoft. Concurrently, in 1979, many computers such as the Osborne 1, with the CP/M operating system, shipped with another version of BASIC developed by Microsoft called MBASIC. Both these simple BASIC interpreters could readily solve simultaneous equations in a manner similar to what FORTRAN did on mainframe computers, but the computers had limited storage capabilities.

When IBM released the IBM Personal Computer (PC) in 1981, it chose to ship it with PC-DOS rather than the then ubiquitous CP/M operating system (popular stories about

Solutions for Soil and Structural Systems using Excel and VBA Programs, First Edition. Robert L. Sogge.
© 2012 John Wiley & Sons, Ltd. Published 2012 by John Wiley & Sons, Ltd.

how this decision was made include a snub by Gary Kildall of Digital Research, who was flying his airplane when a meeting with IBM executives was scheduled). Microsoft's release of a generalized version, MS-DOS, followed in 1982. A BASIC interpreter, created by Microsoft and licensed to IBM, was distributed with IBM PCs for use with these systems. BASICA, a more advanced form of BASIC developed by Microsoft, though still interpretive, was released in 1981. GWBASIC, developed by Microsoft from BASICA, included a compiler and was released in 1982. All these versions required that program lines be numbered. Many engineers who were using FORTRAN programs until the creation of the PC converted them to the BASIC language, despite the latter's numbered lines, slower execution, and minimal storage.

Almost concurrently, in 1981, Microsoft released its version of FORTRAN which had greater storage capabilities for arrays but required a cumbersome (especially with two floppy diskette drives and no hard drive) two-stage process to compile a program. FORTRAN's major strength was its capacity to handle large arrays. When BASIC became more powerful, many programmers switched back to a new variant, QBASIC (more on this later), for ease of compilation and program availability.

Microsoft QuickBASIC was released in August 1985 with updates to version 4.5 in 1988. QuickBASIC, with its vastly increased storage capacity and compilation capabilities, began to displace the use of FORTRAN in engineering program development. QBASIC, a slimmed-down version of QuickBASIC without a compiler, was developed by Microsoft for inclusion with its MS-DOS 5.0 operating system, which was released in 1991. It continued to ship with later MS-DOS versions up until the final stand-alone version 6.22 (1994).

BASIC versions developed for the DOS operating system could still be used with Windows by using a DOS command prompt. Beginning with the release in late 2001 of Windows XP, the GUI (user interface) of DOS was no longer supported. This meant that output screen results could no longer be printed as was possible with Windows 98 and 95, thus beginning a forced shift to the software denoted as Stage II in the next section.

2.1.2 Stage II – Visual BASIC with Windows

The first fully functional Microsoft Windows operating system was Windows version 3.0 released in May 1990. Development of BASIC for use with the Windows operating system, herein denoted as Stage II development, began in 1991 as Microsoft released Visual BASIC (VB) 1.0 in 1991. VB fostered the easy creation of applications using the Windows GUI. Though the language was familiar, the programming interface, UserForm, left the simplicity behind as it tried to encompass and integrate more than engineers required into a new, radically different, screen presentation format. The last version released was VB6 in 1998.

VB is no longer sold or supported by Microsoft. Since VB6 is out of production, the only way of obtaining versions is through used bookstores or from eBay. A time gap, filled in by program releases noted in Stage III, existed before VB was succeeded in 2002 by Microsoft's Visual Basic.NET (VB.NET). The first version was denoted as 7.0, to follow VB6. It was a totally new framework that is not backwards compatible with VB6. Subsequent versions of VB.NET were released in 2003 and 2005.

In October 2008, Microsoft released Microsoft Small BASIC. The program, though simplified, provides an introduction to BASIC programming and works with current operating systems.

2.1.3 Stage III – VBA and Excel with Windows

Stage III programming language development combined the good BASIC language that had been refined up to that point, VB6, with the Excel spreadsheet as a palette for containing the input and output for the VBA programs. Programming was possible without employing the UserForms that had been a cumbersome feature previously. VBA, though essentially the same language as VB6, is specifically associated with the Microsoft Office program suite.

VBA was first included in the 1993 16-bit version of Excel. Excel 95 was the first to make use of a 32-bit operating system. The VBA code in Excel 97 is based on VB5. Excel 97, which is included in Office 97, reintroduced the UserForms programming interface. VBA in Excel 2000 was updated to VB6. VBA version 6.5, which accompanies Excel 2003, although nearly as complete as VB6, of course does not include a compiler. Compiling code reduces its size and decreases the execution time, but since neither the VBA code nor an Excel spreadsheet can be compiled into an executable file, each line of the source code is executed in a slower interpretive mode every time the program is run.

Excel 2010 is the first version to offer both 32- and 64-bit versions as earlier versions came in a 32-bit program that would still work on both 32- and 64-bit operating systems. Really, the two versions are compatible, but the 64-bit version can address a greater amount of RAM.

Both the 32- and 64-bit versions of Microsoft Office contain the new version of Visual Basic for Applications 7.0 (VBA 7). With regard to differences between VBA 6 and 7, previous versions of VBA required the use of the Long data type. VBA 7 contains a specific pointer data type.

Little concerning VBA has changed between the 2003 and 2010 versions of Excel except that the language is easier to implement in later versions. Today, VBA is the common programming language across all of the Microsoft Office programs, and is proprietary to Microsoft. VBA gives any program in the suite the ability to create BASIC language macros. With Excel this combination can utilize the worksheet for pre- and post-processing of data, while the VBA program code acts as the analysis engine. There are many similarities in the language between Stage I and III BASIC. Stage III development essentially preserved the last code development of Stages I and II and in combination with Excel uses a worksheet to display any graphical output. It did provide a LINE command to permit plotting. For engineering, this change was essential as system geometric configurations could not be readily plotted by the chart capabilities of Excel.

2.2 Justification for Using Excel with VBA Macros

The best choice today for engineering calculations is use of the VBA language of Excel along with worksheets. An alternative choice would be to change languages entirely and select one like Java or C++. The languages BASIC, FORTRAN, C++, and Maple contain aspects of both procedural programming languages and an object-oriented approach. Java, MATLAB, MathCAD, and Mathematica are somewhat differently known as array programming languages (APLs).

When one considers the popularity of Excel, the BASIC language, and its similarity to the VBA language, it is evident that the combination provides justification for its use in the curriculum of many undergraduate engineering schools. Additionally, many already have Excel on their systems, so programs created in it can readily be run by other individuals. Moreover, access to sequential data input and output files and transfer of input and output data directly to and from cells in a worksheet, as well as plotting and graphing capabilities, suggest it is a powerful combination analysis tool. As stated by Jerry Pournelle in the March 1994 issue of *Byte Magazine*, ''a competent compiled BASIC programmer can turn out large programs that work much faster than C++ programmers can'' (explanation of context – the program won't run faster, what's faster is getting a working program out and running.) This statement is quite a testimonial for BASIC and still holds true.

2.3 Difference between a Workbook and a VBA Macro

A workbook stores information in a cell of a worksheet. VBA stores information for a constant or variable in its memory in its dimensioned size. In this form it can be manipulated.

There is no incremental or For–Next loop operation possible in a worksheet. For example, in VBA the assignment statement

$$x = 0 \quad x = x + i \qquad \text{Looping structure}$$

cannot be duplicated by worksheet operations. To approximate such an operation in a worksheet, a table must be created containing the various values of x and the values are then checked to see when they exceed a certain value. The workbook **For-Next Demo** demonstrates a For–Next loop that is available in VBA and not in Excel worksheets.

Microsoft Excel worksheets contain math functions that automatically compute typical mathematical operations such as SINE or LOG. Microsoft VBA has functions that are similar to the math functions and perform similar operations. For example, both Excel and VBA have the functions LOG (number, base), which returns the logarithm of a number to the specified base, and LN (number), which returns the natural logarithm of a number.

2.4 VBA Macro Nomenclature

One of the most difficult aspects of learning a programming language is to understand the syntax of the language – describing how to program the statements that comprise a program. A VBA program contains the following objects: Workbooks, Worksheets, Charts, Modules (or Projects) with Macros consisting of Sub or Function Procedures having lines of code consisting of VBA statements. A macro containing Sub or Function Procedures written in the VBA code can be stored in the following objects:

- Worksheet
- Workbook
- Module

A module can have any number of procedures. Project Explorer can be used to see all the objects of a project.

2.5 Generating a Procedure

A VBA macro consisting of procedures developed from lines of code can be created in or copied and pasted into (visible on the Project Explorer screen):

- This Worksheet "Sheet1" under Microsoft Excel Objects of Project Explorer where "Sheet1" is a worksheet name on the bottom tab.
- This "Workbook" under Microsoft Excel Objects shown by Project Explorer.
- A module.

The lines of VBA macro code that are stored in ThisWorkbook can be viewed either through

```
View > Macros > View Macros > edit
```

or through the Developer tab and Project Explorer click

```
Developer > Visual Basic > ThisWorkbook (or Module1)
```

If the Developer tab is not shown, click on

```
File > Options > Customize Ribbon > Under Main Tabs click on
Developer Tools > Macro > Macros
```

Regardless of where the macro is stored, it will have the same name as first executable Sub name.

For the purposes of all VBA code in this text, the choice of storage location for the procedures is arbitrary. It is probably most common to store code in a module. Procedures that are stored in modules can be used or accessed by any other workbook. An association to the workbook containing the module is all that is necessary. There are two types of modules as shown in the VB Editor:

```
Insert > Module or > Class Module
```

This text does not deal with class modules. By saving the macro to a worksheet or workbook, it is associated with the worksheet or workbook and not the module and is not available to other programs. In a later section we will be assigning these macros by their names to a control function in a worksheet.

Examples showing the storage of the VBA code by these various approaches are presented in the following workbooks:

- Workbook **SoilClassUsingVBA** in Chapter 9 shows the Module1 storage approach.
- Workbook **SoilClass-VBA** in Chapter 9 shows the worksheet storage approach.
- Workbook **For-Next Demo** in this chapter shows the workbook storage approach.

Is there a benefit to having a macro stored and run from the same workbook that contains the application data worksheet of Excel as opposed to being stored and run from an individual file?

- The benefit to storing the macro with the application data worksheet as opposed to being stored and run from an individual file is that it is not necessary to link the program or potentially update these links at any time.
- It is preferable, though, to store a macro program as a separate file that then is linked to the data worksheet that uses the macro. This approach permits the development of *the* one macro program and promotes organization of where the last version of the program development is located rather than leaving a scattered array of programs among the various data application worksheets using this macro program on one's computer.

The UserForm operation available from the Tools > Macro > Visual Basic Editor and then Insert > **UserForm** and the associated Toolbox of Controls are not used in this text, nor are they required for any of the code in this textbook. Such forms, which as noted previously are one of the most confusing aspects of VB, are not efficient for large amounts of data. Instead, this text will contain VBA code written in such a manner as to communicate with the user directly through cells of the Excel worksheet, thus using the worksheet as its "UserForm."

With the Project Explorer screen present, code can be entered into a module. Use Insert > Module then Insert > Procedure and enter the code lines. When done, store the code by saving. As stated previously, there are alternatives to where the macro consisting of the assembly of procedures, now in the module, can be stored. The code lines can be selected and copied to the worksheet or the workbook Excel objects.

A module contains a macro that may consist of many procedures. A procedure of the type Sub, Function, or Property is either Public (the default) or Private. Private procedures can be accessed from within that module only and Public procedures can be accessed from anywhere in the project. The lines of code define the Sub or Function Procedure. A module containing Function Procedures appears in the **Flownet** and **LoanProg** programs. Both Sub and Function Procedures are written using the same VBA statement language. A procedure cannot be nested within another procedure.

The code for macros can be generated using either:

- The macro recorder: Tools > Macro > Record New Macro.
- The VBA Editor to enter VBA code: Tools > Macro > Visual Basic Editor > Insert > Procedure.

When using the VBA Editor to create a coded procedure consisting of a function or sub, begin from the Project Explorer screen. Double click on either the Worksheet or Workbook objects or (or use Insert > Procedure) to bring up the screen to enter code. Alternatively use Insert > Module and Insert > Procedure or just begin writing. A procedure that is either Private or Public (the default) begins.

Both Sub and Function Procedure code can be copied from an existing macro shown in other modules into a new program or project in a module shown by Project Explorer and then edited. Copying such as this often simplifies code entry as much of the code in one routine is similar to that in another. This copying can be performed either directly or through the use of an intermediate word processor such as Notepad. Code from older programs, regardless of its language, that may use versions of QuickBASIC, GWBASIC, or BASIC can be copied to the macro module section of Excel from where it can be edited using the VBA Editor. Code can also be transferred from a file containing it using the Insert > File menu in the Visual Basic Editor. Microsoft Visual Basic Help within the Visual Basic Editor is sometimes questionable. As an example, try searching on Module.

2.6 Security Level Required to Open VBA Macros

Viruses often target files containing macros. To permit the opening of workbooks containing macros it is necessary to adjust the default security level in Excel to a lower level. You can adjust the security level in Excel by doing the following:

```
Tools > Macro > Security > medium or low in older versions.
And File > Options > Trust Center > Trust Center Setting
> Macro Settings > Disable with Notification or Enable for Excel 2010.
```

2.7 VBA Code Statements that Differ from Previous BASIC Versions

These are as follows:

- There is no main program and subs with VBA. Instead a macro consists of many sub procedures.
- When entering VBA code it can be confusing if multiple statements are placed on a line using the character ":" or if a statement is continued to another line using the "_" character at the end of one line. The use of either method is discouraged.
- Use of a Go To statement to branch unconditionally out of a loop created by a For–Next unit is also discouraged.
- No Continue statement exists in VBA; instead a line with a number is needed.
- UserForms are not required as was the case with VB6 versions.
- No "Common," labeled block Common, or Common or Dim Shared exists in the Dimension statement. Instead, Dim statements placed outside of the first Sub or Function statement denote variables common to many Sub and Function statements

within a module. This location makes the variable public like Common, meaning public to other procedures (subroutines and functions). Such variables do not need to be repeated within any other sub for that code to have access to its value. Thus, if a variable is identified globally for a module (or at the module level outside subs), then it can be used within all Sub Procedures.

- The Line(x_1, y_1, x_2, y_2) statement of VB6 used for plotting lines has been replaced by an AddLine(x_1, y_1, x_2, y_2) statement in VBA.

2.8 Implementation of VBA Macro Programming

2.8.1 Type and Size Declaration of Variables for Subs and Functions

The **Type** of a variable can be declared in any of the following three ways:

1. **Def**type statement such as DefStr Z.
2. **Type-declaration** character such as %, &, !, #, @, $.
3. **Dim** Z as String, AA as Double, BB as Single – one Dim statement can define variables of the same or different types as well as defining the size of a variable.

2.8.1.1 **Def**type and Variable Accuracy

Deftype can be used to define the type of variable and thereby its accuracy. DefInt, DefLng, DefSng, DefDbl, DefCur, DefStr are statements used to define variables as integer, single precision, double precision, or currency.

A **Def**Type statement must be placed before any other statement at the module level ahead of a Sub Procedure.

The type of a set of variables can be defined using:

- DefInt I-N (This statement declares variables whose names begin with I through N as integer type. Note that it is not necessary that integer types begin with I–N first letters.)
- DefSng A-H
- DefSng O-Y
- DefStr Z.

Each statement must be on one line.

Their declared type can be overridden using a **Dim as** statement, as will be seen later.

2.8.2 Integer Variables

Normal: integer range is -32768 to 32767. The character % appended to a variable name declares the variable's data type as Integer. See Table 2.1 for this and other VBA data type characteristics.

Table 2.1 VBA data type characteristics.

Data type	Character	Digits	Storage size bytes	VBA range limits Positive	VBA range limits Negative
Integer	%	5	2	$2^{15} - 1 = 32,767$	$-2^{15} = -32,768$
Long (integer)	&	10	4	$2^{31} - 1 = 2,147,483,647$	$-2^{31} = -2,147,483,648$
LongLong (integer)[1]	∧	19	8	9,223,372,036,854,775,807	−9,223,372,036,854,775,808
Single (float pt)	!	7	4	3.402823E+38	−1.401298E-45
Double (float pt)	#	15	8	1.79769313486232E308	−4.94065645841247E-324
Currency	@	19	8	922,337,203,685,477.5807	−922,337,203,685,477.5808
String	$	2^{31}	10+String Length	$2^{31} = 2,147,483,648$	
Variant-numbers		15	**16**	up to Double range	therefore don't use Variants
Variant-characters		2^{31}	22+String Length	$2^{31} = 2,147,483,648$	
Decimal[2]		29	12		not for scientific calcs
Date	# enclosure		8	December 31, 9999	January 1, 100
Array(a)			24	plus size of a	
Array(a,b)			28	plus size of a*b	
Array (a,b,c)			32	plus size of a*b*c	

1 byte = 8 bits.
[1]For 64-bit platforms only.
[2]Only used within Variant using CDec function.

Long: integer range is −2 147 483 648 to 2 147 483 647. The character **&** appended to a variable name declares the variable's data type as Long.

Due to the large size of a structural model it is sometimes necessary to make some changes to the standard **PFrame** program. For example, the integer NONB = NODOF* NBW for this model becomes $408 \times 105 = 42\,840$. This value exceeds the integer limit size of 32 767. Therefore it is necessary to Dim NONB as Long. Also, due to the difference in the specific pointer data type between VBA6 and 7 it is necessary to use the function CLng to change one of the variables in the multiplication expression to insure that the arithmetic is carried out using 32-bit numbers rather than 16-bit numbers. If this change is not made, an overflow can occur before the result is converted to a 32-bit number by the type definition Long.

2.8.3 Floating Point Variables

Single-precision variables have seven significant digits of precision. These variable types have a range in value from −3.402823E38 to −1.401298E-45 for negative values and from 1.401298E-45 to 3.402823E38 for positive values. Either DefSng or the character ! appended to a variable name declares the variable's data type as Single.

2.8.4 Double-Precision Variables

Double-precision variables have 15 significant digits of precision and require more memory to store than do single-precision variables. Such variable length is the same as used by Excel in worksheet cell values. Double-precision variables range in value from $-1.79769313486231E308$ to $-4.94065645841247E$-324 for negative values and from $4.94065645841247E$-324 to $1.79769313486232E308$ for positive values. Either DefDbl or the character # appended to a variable name declares the variable's data type as Double.

The workbook **PiTwoSeries** uses the slowly converging series $\pi/4 = 1 - 1/3 + 1/5 - 1/7 + 1/9\ldots$ to compute the value of π. Since the program requires many loops to provide a significant accuracy, it serves as an example of the need for variable types to be declared both as Integer Long and as Floating Point Double Precision. In a For–Next loop, the initial and final values of the counter as well as the counter itself need not be of an integer type and can be defined as double-precision floating point variables. As shown from the results of running this program, to perform the adequate number of loops for accuracy at the 11th digit, the looping index must be greater than even the limits inherent in a Long Integer. A declaration of a double-precision floating point variable type provides 15 digits of precision for the For–Next counter. This type allows the looping counter to exceed the 2 147 483 647 limit of a Long Integer and provide greater accuracy for π. The workbook **PiTwoSeries** contains the Excel worksheet and the VBA code for computing this series.

2.8.5 Currency Variables

Currency variable numbers permit 15 digits to the left of the decimal point and 4 digits to the right. Their range is from $-922\,337\,203\,685\,477.5808$ to $922\,337\,203\,685\,477.5807$. Besides money, this variable is useful for calculations involving fixed-point numbers in which accuracy is important. Either DefCur or the character @ appended to a variable name declares the variable's data type as Currency.

2.8.6 String Variables

The String data type is used for alphanumeric characters. Either DefStr or the character $ appended to a character name declares the variable's data type as String.

2.8.7 Variant Variables

The Variant data type is used for all variables that are not explicitly declared as some other type. A Variant data type can contain any kind of data except fixed-length String data. Variant data types require the most memory of all data types to store their values. The Variant data type has no type-declaration character but can be declared by the statement DefVar.

Appending characters to a variable name to define the type declaration for that variable, such as A# for Double or A% or A&, is not a common practice. Generally, it is better to define these variables through the Dimension statement, as will be seen in subsequent sections.

2.8.8 Declaring Data Types in Sub or Function Procedures

The data types of any of the arguments passed to a Sub or Function Procedure should be declared in or ahead of the calling sub and not in the list of arguments passed. The declaration of variable type of the Function statement itself should be made after the Function statement in the Function macro as follows:

```
Function FNA(M, N) As Single
```

The Function name does not have to be declared or **Dim**ensioned in the calling sub.

2.8.9 Dimensioning Variables

Dim statements are required for specifying the size of arrays. Their other use is to **Def**ine the type of a variable. The best way to define the type of a variable is by using a **Dim** statement. Dim information can be supplied at the beginning of a Sub Procedure or in statements before all Sub Procedures but not in both locations. Variables defined within a sub, whether by Dim statements or not, are not shared with other Sub Procedures. The values for variables Dimensioned in a global defining location, before all subs, are passed to all subs, or in other words are common to all Sub Procedures. Global dimension statements are intended for shared variables and those just mentioned in various procedures. Such a location replaces the old Common statement of FORTRAN. By specifying a global Dim statement or exchanging variables between SUBs ( ), a common or shared storage location is used. This minimizes the amount of storage necessary.

Examples of defining globally the type of a variable are as follows:

Dim I as Integer, N as Long, B as Single, C as Double, E as Currency, Z as String before any of the Sub or Function statements of the module. A separate "as Type" is needed to declare each variable.

"Dim as" statements outside a sub override a DefType statement made outside a sub. If a variable is **Dim**ensioned in statements both outside a sub and within a sub, an error may not be indicated or flagged by VBA but an error in the computational results will occur. As no indication of this error is given in the program execution, it is a very difficult type of error to correct.

In the paired statements

```
Call EQSOL (SS, P, NODOF, NBW)
Sub EQSOL (A, B, NODOF, NBW)
```

a duplicate declaration error is returned if any of the passed parameters are given a Dim as type in the Sub. It is common practice to use Dim AA as double, BB as double, rather than using DefDbl A-B.

2.8.10 Option Explicit Statement

The Option Explicit statement is used at the global or module level, before any procedures (subs), to require that all variables in that module that have not been specified in Dim statements be identified. The use of code with an undeclared variable name will result in an error. The use of the Option Explicit statement prevents the problem of an undeclared variable becoming a Variant type by default or by mistyping the name of an existing variable.

It is good practice to use the Option Explicit statement so that every variable must be declared in a Dim statement before it can be used in the code. It may be painful to identify every variable, but it can be worthwhile. Single-letter variables like A are difficult to search for within code. In fact, it is recommended that all variable names be descriptive in nature and describe the use for the variable.

A statement like DefInt I-N can be used with the Option Explicit statement in effect as long as there are no subs with calling variables like that shown above. For example, the following

Sub XX (A, B, I)

with a DefInt I as integer statement creates a problem.

2.8.11 ReDim Statement

Memory storage requirements can be optimized by using dynamic arrays that are **ReDim**ensioned later in a procedure when their true size becomes apparent. Such an approach is similar to the old Variable dimension statements. It eliminates the need to Dim them with large sizes to accommodate any general large-scale data requirement. Variables to be ReDimensioned at a later point in the procedure should be specified using ( ) in the Dim statement and then later specifying the sizes of the variables with either single or dual dimensions using the **ReDim** statement when their size is known. A ReDim or variable dimension approach can be used to specify subscripted variable sizes since the program is not compiled.

If the Dim statement has been used to define the type of variable then the same definition should be used as the variable is ReDimensioned later in the procedure, that is, ReDim S(a) As Double.

2.8.12 Sub Procedure

In the Sub Procedure, subs can be accessed by the statements GoSub and Return, or Call and Return. Subs identified by a line number and a Return statement within a sub can be created and called using GoSub LineNumber and Return. When accessing a sub using the GoSub statement, all variable values are transferred since code statements with this LineNumber would be within the same Sub Procedure.

A more structured alternative and thus desirable approach to using GoSub statements would be to create separate Sub Procedures that can be accessed by name rather than line number, using a Call statement. Two End Sub lines cannot be in a procedure. Instead, Exit Sub should be used. Such procedures can have the same line numbers and the same variable names as in other procedures. The variable values will not be shared unless they are in the global Dim statement placed before the first procedure.

The storage of passed variable values to Sub and Function Procedures is specified in the first Dim statements for those variables. Transfer of variable values to called subs is shown in the following statements:

```
Call EQSOL (SS, P, NODOF, NBW)
Sub EQSOL (A, B, NODOF, NBW)
```

Both SS and P are Dimensioned and Defined outside the subs. Both A and B occupy the same storage location as SS and P and therefore do not require additional Dimensioning of storage space.

The statements

- DefDbl A-H, O-Y
- DefInt I-N
- DefStr Z

in a module outside of all subs will not work if there are subs like

```
Call EQSOL (SS, P, NODOF, NBW)
```

that are passing arguments. (See program **HSpace** and **PFrame**.) Instead the variable type should be declared using Dim statements.

Line numbers are not shared between subs. Therefore the same line numbers can exist within different subs

2.8.13 Function Procedure

A Function Procedure can be created using VBA statements. It acts like a sub in that all variables used in the lines of code must be defined in the referring procedure in a Dim statement. An example of a Function statement is shown in program **Flownet** in Chapter 15.

Variable values can be passed to the Function Procedure by listing them after the name. The actual name of the function can be set in the procedure to the value that needs to be returned from the VBA Procedure to the worksheet cell containing the Function name. The value in that cell will not change until the value of any variable passed through to the Function is changed. The value is returned from the Function Procedure in its name. The data type of the name passed from the procedure is declared as Single after the Function statement.

2.9 Inputting Data to a VBA Procedure

Data consisting of values of variables for a VBA macros program can be input by any of the following methods:

1. Values in the cells of the worksheet.
2. Sequential data text input file.
3. **InputBox**es that appear after the program is initiated.

Or some combination of the above. We will examine these separately in the following sections.

2.9.1 In Worksheet Cells

The transfer of the value in the cell of a worksheet to a VBA program variable can be accomplished by one of the following equivalent statements:

```
NNODES = Cells(13,3) (row=13, column=3)
NNODES = [C13] (column C, row 13)
NNODES = Range ("C13").Value
```

All three of the above statements are identical and will transfer the value in the specified cell to the variable NNODES in the VBA macro. The entry [C13] can be used wherever this value is needed in a formula in any cell. For organizational purposes it is best to read the value of [C13] into a named variable and reference the variable when required.

The advantage of using the Cells(4,3) notation over the [C4] notation is that variables can be assigned, Cells(X,Y), to describe the cell data locations. Examples of using code lines as described to automatically sequence the locations of input and output data are shown in the macro program **PFrame**. The use of values that index the (row, column) coefficients of the Cells statement, within a VBA macro For–Next loop, can automate the process of transferring values either into or from a worksheet. Programs that take this approach contain statements in the code of VBA procedures like

```
X = Cells(IROW, 14)
```

Statements out of **PFrame** code are

```
IROW = Cells(8, 3)
ZTitle = Cells(IROW, 1)
Print #2, ZTitle

Print #2, "No of: Jts", "Members", "Matls", "Suptd Jts"
IROW = IROW + 2
NNODES = Cells(IROW, 1)
```

The beginning input (and output) code row line numbers are passed to the VBA macro from a specific (row, column) location in the referring worksheet. This type of input eliminates the necessity to alter any code line in the VBA macro. Typically, any changes in the row or column locations of cells in the worksheet would not automatically transfer to corresponding changes in the macro.

2.9.1.1 Using Worksheet to Pre-process Input Data

By creating input data for the VBA macro directly in the worksheet cells below cells labeled with "headings," the Excel worksheet in effect performs the function that pre-processor programs did in creating "batch" input data files. The Excel worksheet will have headings, and in the rows and columns off to the side a description of the data and supporting computations can be made in these adjacent cells to develop values for the input cells. This approach constitutes the best method of inputting data to a macro program.

In summary, the advantages of inputting data to the VBA macro directly from cells in a worksheet are as follows:

- It is not necessary to use a text editor such as Notepad outside of Excel.
- The worksheet acts as a pre-processor through its headings, or as a template for what data needs to be entered, handling the pre-processing of input data to create the "batch-type" input data file.
- Any editing of code in the macro or the continual typing to an InputBox for data runs is eliminated.
- The cells in the area to the right and outside of the input data cells can be used as a scratch sheet for calculations of any kind, as they are ignored on input.

A minor disadvantage of this input data approach is:

- the requirement to store each workbook, together with the different individual input data values in the cells of the worksheets, as a separately labeled project data file.

2.9.2 From a Sequential Data File

A sequential data file consisting of the values of variables required in a VBA Procedure can be created as a text file using a text editor such as Notepad or WordPad using the proper filename type. The file can be read using an Input # statement in a code line of a VBA Sub Procedure.

To run Notepad, identify the file using Windows Explorer or the Search feature of Windows available from Start Search, **Open With** Notepad (N), selected from the list offered. By entering N, Notepad is accessed as there are no other choices beginning with N within the selection. Alternatively, use Start > Run > Notepad to access the text editor.

Enter the data, consisting of numbers or letters, into the file by typing in fields one after another separated by commas, one or more blanks, or a carriage return. Entering such data below appropriate "title" headings makes it easier to organize the data. The input of "title" headings into the VBA program is performed using a Line Input statement.

Save the input data file as follows:

```
drive\location:    c:\My Documents\ENGR docs\File location\
name:         "Filename.txt"
```

Change ENGR docs to the name of the file under My Documents where you stored (put) this file after opening it in Excel and remember to enter the "\" at the end.

Save as the type. txt using ANSI encoding. The data file created by Notepad is a text file with the extension. txt. It is important to enter the. txt extension in the filename when using a filename input to a macro. The Excel spreadsheets and the associated input and output data files for attached macros need not be stored within the same folder.

An example of using an input disk drive\location and input filename of a sequential data text input file to a VBA Procedure from the cells of an Excel worksheet follows. The VBA Procedure must contain a line of code similar to

```
ZD = Cells(4,3) (row, column notation) or alternatively [c4]
or [C4] (column, row notation)
```

and Cell(4,3) of the worksheet must contain an entry similar to

```
"c:\My Documents\ENGR Docs\"File location"\
```

Obviously, the correct folder structure as above would need to be created within your computer's file structure. The "Instructions for Use of Program Disk" file that is supplied with the disk provides further information on this subject.

Similarly, for the input filename, the VBA Procedure must contain a line of code similar to

```
ZF = Cells(5,3) or alternatively ZF = [c5]
```

and Cell(5,3) of the worksheet must contain an entry similar to

```
"filename.txt"
```

Statements of **Flownet** code that strip off the. txt extension and add OUT to the filename are

```
ZD = [D4]
ZF = Cells(5, 4)
ZDF = ZD + ZF
Open ZDF For Input As #1
Length = Len(ZF) - 4
ZFF = Mid(ZF, 1, Length)
ZFO = ZFF + "OUT"
ZDO = Cells(34, 4)
ZDF = ZDO + ZFO
Open ZDF For Output As #2
Print #2, "*** Program  Flownet  by R L Sogge ***"
```

```
Print #2, "Input Drive\Location, Input Filename = "; ZD; ZF
Print #2, "Output Drive\Location, Output Filename = "; ZDO; ZFO
```

The advantages of inputting data using a sequential data file are as follows:

- You have a labeled file of the data for each project you work on.
- The same workbook will work with many different input data files.
- By constructing the file with headings, input is essentially as easy as from a worksheet and no pre-processing program is need to create a "batch-type" input data file.
- The VBA code is more general and does not need to be varied for cell location (although this is done automatically within the VBA code using the approach taken in Section 2.9.1).

2.9.2.1 A Note Concerning Macro Code Statements for Inputting Data from a Sequential File

When inputting data for variables of a VBA Procedure from a sequential file, the following statement cannot be used since the first input value of NPT will not be transferred to the subscripted variables:

```
For I = 1 to N
Input #1, NPT, X(NPT), Y(NPT)
Next I
```

Such a statement is for unordered (non-sequentially numbered) data input. It is good practice to input joint and member numbers in sequential order.

Instead the above should be written as

```
For I = 1 to N
Input #1, NPT
Input #1, X(NPT), Y(NPT)
Next I
```

In the data file the values of NPT and X(NPT), Y(NPT) could both be on the "same line," or they could be on separate lines of the sequential data input file. If the value for X(NPT) is on the "next line" after NPT, a carriage return is necessary after the NPT value in the data file.

Such an input data file structure works also for non-sequentially ordered number data. Examples of the latter would be boundary condition input in programs **Flownet** or **PFrame** shown by the following statements:

```
For I = 1 to NFBC
Input #1, N
Input #1, B(N)
Next I
```

Placing two statements on one line using the colon (:) is equivalent to the above, yielding:

```
Input #1, N : Input #1, B(N)
```

In the data file, N and B(N) could both be on the same line or they could be on separate lines of the sequential data input file. The reading of the input file will remain on the same line in the data file until a carriage return is encountered (this is not like the original FORTRAN card reader statements where the reader would drop to a new line (card) every time the Input (Read) statement was encountered).

2.9.3 From Input Boxes

An example using input boxes for input file location and name is as follows.
For the input drive\location,

```
ZD = InputBox("Drive", "INPUT", "C:\Users\Robert\My
Documents\DATA FILES\"File location"\)
```

For the input filename = ("filename.txt") where filename.txt is any valid filename

```
ZF = InputBox("Filename", "INPUT", "filename.txt")
```

The InputBox approach forces the user to look at the variable names specified for these parameters. This approach is not used in the programs in this text.

2.10 Output Data from a VBA Procedure

Similar to input data, output data from a VBA Procedure can be directed to cells of a worksheet, a data file, or to a line plot in an area of the worksheet.

2.10.1 Output to Worksheet Cells

Data generated by the VBA code can be transferred to cells in the worksheet, for easy viewing and charting of the data, by using the following statement in the VBA macro:

```
Cells(IROW, 3) = SS(J+1)  where array is in (row, column) notation
```

For data manipulation and plotting it is best to send output data to the cells of a worksheet. Once there, the data can be manipulated by Excel, or even charted using Excel Charts. No separate post-processing program is needed. It is not necessary to ensure that the cell range is cleared between runs, since output data from previous input data runs will just be overwritten with new output data.

Similar to specifying the beginning input code row line number, the beginning output line number is input to the VBA Procedure from a specific (row, column) location in the referring worksheet. This input eliminates the necessity to alter any code line in the VBA macro.

2.10.2 Output to a Sequential Data File

Data generated by a VBA program can be transferred to a sequential output data file by defining the drive\location\name for the output data file in the worksheet cells for input to the VBA program. Again, as with an input file, the naming of the data file information (drive\location and name) for the macro is entered via worksheet cells:

```
drive\location: C:\Users\Robert\My Documents\
DATA FILES\Curve Fit\
name: 'Filename'
```

   This information should be located in the appropriate worksheet cell location, for example, [c4], [c5].

   Programs in this text automatically create an output data file with the name "FilenameOUT" and put it at the drive\location specified. A sequential file output from a VBA macro does not have any extension unless one is specified as above. The output file created can readily be viewed by using Windows Explorer to locate the file and then by using Notepad to open and view the output file. If saved, after being opened using Notepad, such a file is then denoted as a. txt file in ANSI encoding.

   A sequential output data file is opened and output data is written to it using the Print # statements. Print # statements are used rather than Write # statements, as the latter insert commas between items and quotation marks around strings as they are written to a file. When data is output from a VBA macro Procedure to a sequential data file, since there is not a Print Using statement, the best way to format the variable of the data output would be to use the following statement:

```
Print #2, Format (AA(6), "0.00E+00 ");
```

To separate values, put spaces at the end, not at the beginning.

   A comma (,) in the end position prints the next output value in the next 14 character-wide print zone on the same output line.

   A semi-colon (;) in the end position prints the next output value according to the given format immediately after the last value on the same output line.

   For all Excel workbooks that refer to an assigned VBA macro, both the input and output data are written by a sub of the VBA macro to a sequential data output file. Such output of the input data to an output data file, line by line, as data is read into a procedure from cells in the worksheet, has always been considered good coding practice. Then, on a failed execution due to bad data, the computation stage of the VBA macro and line location of input file data errors can be determined from the last input data written to the output file. This practice works better for input from a sequential data file than from cells of a worksheet, as there is a lag present in transferring output data to cells. In other words, execution of the VBA macro ceases when the file operation is reading the data file, and this may be well before previous input data has been sent to the cells of the worksheet.

2.11 Running a Macro

2.11.1 Using a Start Button

A VBA macro program can be initiated or run from the data worksheet without writing or modifying any macro code in VBA. To run a VBA Sub Procedure, it is best to use a start button that has been created using a Form Command Button in an Excel worksheet of a workbook containing the Sub Procedure.

To create a "Start" button, go to View > Toolbars > Forms in the worksheet menu bar. Place the cursor where you want the button to appear. Double click on the Button icon.

This procedure is the same as using Forms Controls through the Tools > Customize > Toolbars > Forms.

In Excel 2010 the easiest way to create a "start button" that when clicked on will initiate a macro is to insert a graphic object onto the worksheet by and then making it a control button by assigning (linking it to) a macro. The most convenient graphic object is usually a rectangle. So the procedure is to:

```
Using Insert > Illustrations > Shapes > Rectangle
```

and then right clicking on the object to assign it to a specific macro. This assignment can best be done by having the macro that is being assigned open at the time of assignment.

Start program:

Start PFrame

A major advantage of creating a "Start" button from the Forms toolbar as described to control an event, in this case the execution of a VBA macro in a workbook, is that such use does not require modification of, nor does it change, any VBA macro code lines in the referred to Sub Procedure. An example of the use of a "Start" button is shown in the workbook **PiTwoSeries**.

The question is often asked, where must the workbook containing an Excel macro be stored for a Start button in an Excel workbook to find and open it? The following procedure for assigning the Start button to a macro addresses that question:

1. Open the workbook containing the macro that is to be run (the macro can be stored in any workbook).
2. In the workbook that utilizes this macro right click on the Start button and assign it to the desired Sub name out of those listed from the list of All Open Workbooks.

After the Excel workbook is executed once, Excel will create, automatically, the path (link) to the assigned macro workbook so it can locate the proper workbook and open it. This will be done even if the workbook was not open. Once developed, this path cannot be edited. This procedure eliminates having to enter a long path to the workbook containing the macro sub to be executed. This linking of the button to a workbook is really no different than when curves in a chart are linked to referenced data in a table in another workbook.

In order to store a macro program in the 2010 version of an Excel workbook, the workbook must be saved as a Macro-Enabled type. A workbook that is merely linked for execution purposes to another workbook containing a macro need not be saved as a Macro-Enabled type.

The best approach to keep the VBA macro program code independent of the data being input from a worksheet is to use the Start button, the bottom tab name, and the beginning row numbers for the data input and output lines. The VBA macro can then be used with any worksheet containing data, and the data can be manipulated in the worksheet rather than by altering the VBA macro. This input data need not necessarily start in the same cells as the start location is input to the program.

2.11.2 Alternative Start Methods

Alternatively, without a Start button, the macro can be run in the VBA Editor. Bring up Project Explorer by selecting the following from the Microsoft VB screen:

```
View > Project Explorer
```

Select the Excel Object (Worksheet, Workbook, or Module) with the VBA code and then

```
View > Code or by double clicking on the Visual Basic
Editor, the code appears.
```

Enter **Run** with the cursor on the Sub that initiates the VBA macro program.

Function macros are not shown in this Tools > Macro > Macros sequence since they cannot be run by themselves.

Although macros can be run using a couple of methods mentioned, only the Start button, created as described above, will be used in the programs of this text, since it is the easiest approach. While running a VBA program, Windows Explorer, or the Search feature of Windows available from Start Search, should be left open. Notepad can then be used to access and edit the input data file and open and view the most recent output data file.

With some programs within this text, and specifically **PFrame**, which is associated to a specific worksheet for plotting, in order to keep the VBA program code unchanged during its use in any application, the name tab at the bottom of the worksheet containing the Start button that calls the **PFrame** program should, for consistency, be named "PFrame" (not case sensitive). Such naming will then be consistent with the name used in the macro Sub PLOT code line Sheets ("Name").select.

2.12 Code Debugging

Debugging of the code can be done by using the Debug shortcut on the Visual Basic Editor menu bar. Execution of a macro program that is running in an infinite loop or from which an exit is desired can be stopped by pressing the Break key or the Ctrl and Break keys simultaneously. Just by leaving off #2, in the following print statement:

```
Print M, FMIN(M,1)
```

an error indicated by the statement,

The error Object doesn't support this property or method (Error 438) can occur without showing you the line where the error occurs.

2.13 Charting in a Worksheet

Charting of output data from a VBA Sub Procedure is best done in the worksheet by filling cells of the worksheet with the data from the VBA macro. This output data can be graphically displayed by using the Excel chart options. The chart type used exclusively in this book is XY (Scatter).

The charting functionality available with Excel is unable to plot non-connected or connected line data by using one data series. For example, Excel charts cannot take a set of data and connect the endpoints to create a plot of a simple two-story frame. Instead, as many data series must be created as there are lines. This restriction within Excel charts greatly limits its graphical plotting capabilities for structural configurations.

2.14 Line Plots in a Worksheet

A VBA sub can plot lines directly to a named worksheet. First the name of the worksheet to which the plot should go needs to be defined in the routine. For example,

```
Sheets("PFrame").Select
Set myDocument = Worksheets(1)
```

Plotting of lines by a VBA sub directly to the named worksheet is performed using a powerful statement called AddLine. The VBA Plot Sub Procedure would contain the following statements:

```
With myDocument.Shapes.AddLine(X(IPP), Y(IPP), X(IQQ), Y(IQQ)).Line
.DashStyle = msoLineDashDotDot
End With
```

Using the above code lines for each member and their joint endpoints of a frame develops the plot of the structural configuration. Coordinates for worksheet–screen plots are measured positive down and to the right. Scaling of the coordinates is required to fit the plot within the selected worksheet output area. The worksheet area depends on the screen size and resolution.

For a 10 in plot width $\times$ 6 in plot height area and a screen resolution of 640×480 (horz $\times$ vert), then

On screen horz: 100 points $= 133$ plotted points $= 2.09$ in (47.8 pts/in on screen)

On screen vert: 100 points $= 100$ plotted points $= 1.7$ in (57.1 pts/in on screen)

For such a ratio of worksheet inches to plotted points the following statements will scale the line endpoints.

The scale factor is the greatest of

```
SFAC = XD / PlotWid
If YD/PlotHt > SFAC Then SFAC = YD/PlotHt

For I = 1 To NNODES
XP(I) = (X(I) - XMIN) / SFAC * 47.8 + XORIG * 47.8
YP(I) = (-Y(I) - YMIN) / SFAC * 57.1 + YORIG * 57.1
Next I
```

To remove the plot on the worksheet use the select arrow in the Home > Editing > Select menu, then delete.

It would be beneficial to store the examples in separate worksheets all contained in one workbook labeled Chapter 4-Examples. This approach is not possible for the reason that the statements:

```
Sheets("PFrame").Select
Set myDocument = Worksheet(1)
```

that are required for the PLOT macro of program PFrame cause a problem with the requirement that one cannot rename worksheets to the same name as another worksheet or a workbook referenced by Visual Basic. It would be necessary to modify these statements in the PLOT routine of PFrame each time removing the independent nature of this macro. Therefore each example is stored in a separate workbook with the worksheet labeled "PFrame".

2.15 Macro Sub Program Showing Output to Worksheet

The output of graphical data from a VBA macro to a worksheet is shown in the four Sub Procedures in the workbook entitled **LogoOutput**.

The first macro Sub Procedure, Logo, outputs a graphical display consisting of a company name in the cells of the worksheet in the form of a sine curve. Do loops are inserted to slow the process such that the output has some rhythm to it as it is created.

Long integer numbers are required in some cases. A DoEvents statement is required to make this program run smoother with most processors (see below).

The Sub Procedure ChartAnimator shows how a VBA macro can change the data in the cells on which a chart is based and in essence animate the chart. In this specific program the VBA program creates a loop and the spreadsheet charts create different sine curves.

With the faster CPUs available now, the previous command may not be finished before processing has proceeded to the next command. A DoEvents statement will allow processing to complete a command before proceeding to the next. Without such a delay statement the charting in the spreadsheet will not be processed before subsequent statements are started and no change appears in the chart.

The Sub Procedure PlotSine plots a sine curve as a series of lines to a section of the worksheet where the plot is centered based on its extents.

The Sub Procedure Plot does the same for a superimposed triangle and circle.

2.16 Computer Hardware/Software Requirements

All workbooks and VBA macro programs were developed for Excel 2010. A PC with any operating system capable of running Excel 2003, or the later 2007 and 2010 editions, can run the spreadsheets and programs.

The software has been tested and checked on the following hardware/software combinations:

- Windows XP and Excel 2003, Dell Inspiron 5100 Laptop Intel Pentium 4 CPU, 2.66 GHz, 1 GB RAM
- Windows XP and Excel 2007, Dell 11 Netbook with Intel Atom (N270) CPU, 1.6 GHz, 1 GB RAM
- Windows 7 Professional and Excel 2010, Dell Inspiron 14R (N1440) Laptop with Intel Core i5 CPU, 2.4 GHz, 4 GB RAM.

As none of the programs is video or graphics intensive, it is not necessary that the graphics processing unit (GPU) be discrete or especially powerful. On the computers tested, the GPU was integrated on the motherboard (as opposed to discrete, and separate).

2.16.1 Memory Requirements

As stated in the Excel Visual Basic Help menu: "The maximum size of an array varies, based on your operating system and how much memory is available. Using an array that exceeds the amount of RAM available on your system is slower because the data must be read from and written to disk." The memory that can be created outside of the physical RAM chip memory is called virtual and consists of a file on the computer's hard drive that is 1.5–3 times the RAM size. The reader is referred to a Windows text for information on managing these files.

In Excel 2003 the =INFO("memavail") statement gives the available (free) memory in bytes to run the program. This statement returns the same number as the following statement in a VBA Procedure:

```
MsgBox "Excel has" "Application.MemoryFree" "bytes Free"
Total-Used = Free in VBA using MsgBox
Total-Used = Avail in Worksheet using =INFO.
```

These statements are unavailable in Excel 2007 and 2010 and the VBA associated with them. In Excel 2003 a couple of items appear to be in error with the =INFO("memavail") statement. First, the value returned is in kB (kilobytes) and not bytes. Second, as this value stays constant regardless of the program or the computer's RAM size, it appears that it is more a statement of the limit on the working memory used in processes by the Excel 2003 programs.

Excel 2003 processes have a Windows XP memory limit of 1 GB that includes the virtual memory limit. In Excel 2007 this limit is 2 GB. The memory usage can also be obtained by Windows Task Manager, started by hitting the Ctrl, Alt, and Delete keys simultaneously, starting Task Manager, and then viewing its Performance tab. The memory stated by the statements mentioned above is not in agreement with those reported in this Performance tab.

For the program **PFrame**, presented in Chapter 4, a translation of the storage requirements for the various variable types to the total memory required (all related to the number of joints) and the memory limits available is done in the workbook **PFrameMemSpace**.

The critical operation within **PFrame** that is limited by the memory are the two statements:

```
NONB = NODOF * NBW - limited by Integer size
RedimSS(NONB) - limited by double precision storage of
   floating point variable.
```

A Long integer type essentially removes the first Integer size limitation.

As the storage of the Double Precision variable SS takes up the majority of the memory capacity in program PFrame, the running of the program to that point, as in done in workbook **PFrameMemSpace** shows how large this variable can be dimensioned before there is a memory overflow indicated by the statement:

```
Run-time error 7 - out of memory
```

occurs at that line in the code. The limit on SS size appear to be SS(70,000,000). This approximately 0.5 GB limit is reached when the number of joints exceeds approximately 3900.

2.16.2 Processing Speed

The processing of VBA macro calculations is very CPU intensive. Windows Task Manager shows the memory usage of the various processes that the computer is

Table 2.2   Time results for program Pi Excel 2003, Windows XP, Intel Pentium 4 clocked at 2.66 GHz, 1 GB RAM.

Number of terms in series	Computed value of π	Execution time (min:s)
32 767 (Integer)	3.141 562 136 011 68	0:00
100 000 000	3.141 592 643 589 37	0:14
500 000 000	3.141 592 651 589 41	1:09
1 000 000 000	3.141 592 653 589 79	2:20
2 147 483 647 (Long integer)	3.141 592 653 122 16	5:00
4 000 000 000	3.141 592 653 337 77	9:14
10 000 000 000	3.141 592 653 485 72	22:43

Value of π accurate to 15 digits is 3.141 592 653 589 79.

For comparison purposes, the time for the 100 000 000-term computation using Excel 2007, Windows Vista, Intel Core 2 Duo CPU, 2.1 GHz, 4 GB RAM is 4 seconds; and Excel 2010, Windows 7, Intel Core i5 CPU, 2.4 GHz, 4 GB RAM is between 4 and 5 seconds.

executing and the performance as measured by CPU usage per % of total capacity. During execution of the workbook VBA program **PiTwoSeries** it is possible that 100% of the CPU will be utilized, and during this processing the CPU may not be free to perform other tasks. In running such a program, the benefit of a dual-core or even quad-core CPU becomes evident. With such a system, a process that is normally shared by with other concurrent processes can be assigned to one of the cores, allowing other tasks to be conducted while the calculations are proceeding.

The Time function in VBA retrieves the system time at the beginning and end of the macro program execution. In program **PiTwoSeries** time values are sent to the worksheet cells by the statements [c11] = Time, and [d11] = Time, where their computed difference gives the execution time. Computation times for various terms in the π series are given in Table 2.2. It is from the long execution times that programs like this take, when their processors can make many millions (mega) of operations per second, that the true magnitude of a billion (1×10^9) becomes evident.

The speed with which a computer processes tasks is dependent more on the speed of the CPU than on the amount of RAM. For computers using Windows XP with a 1.6 GHz CPU and 1024 MB (1 GB) of RAM installed, neither speed nor memory size will be a problem when executing any of the programs presented in this text. None will take longer than a few seconds of execution time to complete.

   Related Workbooks on DVD

For-Next Demo with VBA macro stored in this workbook.
PiTwoSeries with VBA macro stored in Module1.
PFrameMemSpace.
LogoOutput with four VBA macro subs stored in Module1.

Further Readings

Bullen, S., Bovey, R., and Green, J. (2005) *Professional Excel Development*, Addison-Wesley.
Shepherd, R. (2010) *Excel 2007 VBA Macro Programming*, McGraw-Hill.
Walkenbach, J. (2004) *Excel 2003 Power Programming with VBA*, John Wiley & Sons, Inc.

Historical (Ascending Date Order)

McCracken, D.D. (1965) *A Guide to FORTRAN IV Programming*, John Wiley & Sons, Ltd.
Bowles, J.E. (1974) *Analytical and Computer Methods in Foundation Engineering*, McGraw-Hill (FORTRAN language programs).
Bowles, J.E. (1977) *Foundation Analysis and Design*, 2nd edn, McGraw-Hill.
Microsoft (1979) MBASIC Microsoft BASIC-80 Reference Manual (distributed with Osborne Computer).
Poole, L. and Borchers, M. (1979) *Some Common BASIC Programs*, 3rd edn, Osborne/McGraw-Hill.
IBM (1981) BASIC by Microsoft (distributed with IBM PCs).
Miller, A.R. (1981) *BASIC Programs for Scientists and Engineers*, Sybex.
Jenkins, W.M., Coulthard, J.M., and de Jusus, G.C. (1983) *BASIC Computing for Civil Engineers*, Von Nostrand Reinhold.
Microsoft (1984) Microsoft FORTRAN Compiler (for MS-DOS), User's Guide 165pp, Reference Manual.
Gottfried, B.S. (1986) *Theory and Problems of Programming with BASIC*, Schaum's Ouline Series, 3rd edn, McGraw-Hill.
Microsoft (1988) Microsoft QuickBASIC BASIC Language Reference, Version 4.5.
Microsoft (1990) Microsoft QuickBASIC, Version 4.5, Learning to Use Microsoft QuickBASIC 322pp, Programming in BASIC.

Part Two
Structures

"Since the mathematicians have invaded the theory of relativity, I do not understand it myself anymore."

– Albert Einstein

Part Two
Structures

3

Finite Element Method – The Theory

3.1 Theory

Most systems composed of soil and a structure are highly indeterminate. Often a structural problem is made determinate by:

- defining the failure or collapse mechanism;
- specifying a deformation or a stress pattern.

In the past structural systems having only a few degrees of indeterminacy were solved by employing structural analysis methods such as moment distribution, conjugate beam, or unit dummy load (Norris and Wilbur, 1960). For highly indeterminate systems it is best to set up and solve the equations of elasticity (Zienkiewicz, 1971; Przemieniecki, 1968; Cook, Malkus, and Plesha, 1989; Buchanan, 1995), namely:

- Equilibrium
- Force–deformation (stress–strain)
- Deformation compatibility.

Note that, for references, this book uses basic, well-established procedures; therefore the reference list relies on publications that first established these procedures. The text does not incorporate the latest journal articles on the various topics, but it does incorporate the latest programming techniques in implementing these procedures.

Solutions for Soil and Structural Systems using Excel and VBA Programs, First Edition. Robert L. Sogge.
© 2012 John Wiley & Sons, Ltd. Published 2012 by John Wiley & Sons, Ltd.

3.2 Developing the Element Stiffness Matrix

The finite element method or displacement method of structural analysis involves setting up and solving the following three sets of equations associated with any indeterminate system.

3.2.1 Equilibrium

The equilibrium equations for any element (member) are

$$\{P\}_{6\times1} = [A]_{6\times6}\{F\}_{6\times1}$$

relating the externally applied forces, $\{P\}$, in a global coordinate system to the internal member forces, $\{F\}$ oriented in a local coordinate system. The $[A]$ array is basically a coordinate transformation that will be used later to convert a stiffness matrix from the local (element) coordinate system to the global (system) coordinate system. It is equivalent to the direction cosines for the element or

$$\begin{vmatrix} \cos\theta & -\sin\theta \\ \sin\theta & \cos\theta \end{vmatrix}$$

which determinant is one.

The $[A]$ matrix is strictly a coordinate transformation matrix and does not include the effects of sidesway. The stiffness array for the element in local coordinates includes all effects of sidesway by including the shear forces, F_2 and F_5.

The local and global coordinate systems and the associated member and joint forces are shown in Figure 3.1. Values for the elements of this array are presented in Figure 3.2.

The symbol $[A]_{6\times6}$ is used in matrix algebra to denote the equilibrium array that is a matrix of coefficients relating the $\{P\}$ and $\{F\}$ variables. The subscripts denote first the number of rows and next the number of columns of the matrix $[A]$. Matrix operations such as addition and subtraction can be performed on matrices of similar order (equal number of rows and columns). Two matrices can be multiplied when the number of columns of the pre-multiplier equals the number of rows of the post-multiplier.

A determinate system is one in which the member forces $\{F\}$ can be computed directly from the applied joint forces, P, using equilibrium equations only. With determinate solution methods the stress–strain and strain compatibility conditions are incorporated by approximating the structural deformation and stress distribution.

In a statically determinate system the equilibrium equations alone are sufficient to solve for the forces in the system; the stresses are independent of E. In a statically indeterminate structure the values of E and v in the force deformation equations are needed for computing the forces in the structure. In such a structure, the path that

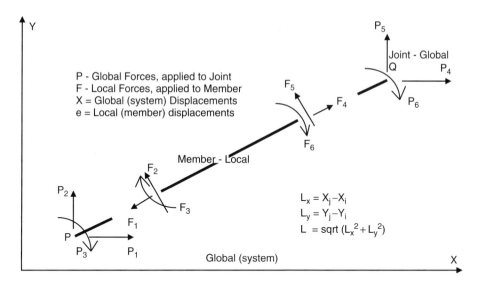

Figure 3.1 Positive global and local coordinate system.

$$
\begin{bmatrix} P_1 \\ P_2 \\ P_3 \\ P_4 \\ P_5 \\ P_6 \end{bmatrix}
=
\begin{bmatrix}
L_x/L & -L_y/L & & & & \\
L_y/L & L_x/L & & & & \\
 & & 1 & & & \\
\hdashline
 & & & L_x/L & -L_y/L & \\
 & & & L_y/L & L_x/L & \\
 & & & & & 1
\end{bmatrix}
*
\begin{bmatrix} F_1 \\ F_2 \\ F_3 \\ F_4 \\ F_5 \\ F_6 \end{bmatrix}
$$

Figure 3.2 $\{P\} = (A)\{F\}$.

the load takes to the supports and the forces in the structure are dependent on the distribution of E throughout the structure, but not on the magnitude of E.

For no special conditions such as hinges within the structure, a determinate frame exists if

$$3 \times \text{No. of joints} - \text{No. of support directions} = 3 \times \text{No. of members}$$

If the left side of the above equation is less than the right side then the structure is statically indeterminate. For trusses with two degrees of freedom at a joint, the equation determining statically determinate structures is

$$2 \times \text{No. of joints} - \text{No. of support directions} = \text{No. of members}$$

$$
\begin{bmatrix} F_1 \\ F_2 \\ F_3 \\ F_4 \\ F_5 \\ F_6 \end{bmatrix}
=
\begin{bmatrix}
AE/L & 0 & 0 & -AE/L & 0 & 0 \\
0 & 12EI/L^3 & -6EI/L^2 & 0 & -12EI/L^3 & -6EI/L^2 \\
0 & -6EI/L^2 & 4EI/L & 0 & 6EI/L^2 & 2EI/L \\
-AE/L & 0 & 0 & AE/L & 0 & 0 \\
0 & -12EI/L^3 & 6EI/L^2 & 0 & 12EI/L^3 & 6EI/L^2 \\
0 & -6EI/L^2 & 2EI/L & 0 & 6EI/L^2 & 4EI/L
\end{bmatrix}
*
\begin{bmatrix} e_1 \\ e_2 \\ e_3 \\ e_4 \\ e_5 \\ e_6 \end{bmatrix}
$$

develop column by column
for column 1 set $e_1 = 1$, $e_2 - e_6 = 0$
for column 2 set $e_2 = 1$, e_1, $e_3 - e_6 = 0$

Figure 3.3   Member stiffness array $\{F\} = (S)\{e\}$ in local coordinates.

$$
\begin{bmatrix} F_1 \\ F_2 \\ F_3 \\ F_4 \\ F_5 \\ F_6 \end{bmatrix}
=
\begin{bmatrix}
AE/L & 0 & 0 & -AE/L & 0 & 0 \\
0 & 3EI/L^3 & 0 & 0 & -3EI/L^3 & -3EI/L^2 \\
0 & 0 & 0 & 0 & 0 & 0 \\
-AE/L & 0 & 0 & AE/L & 0 & 0 \\
0 & -3EI/L^3 & 0 & 0 & 3EI/L^3 & 3EI/L^2 \\
0 & -3EI/L^2 & 0 & 0 & 3EI/L^2 & 3EI/L
\end{bmatrix}
*
\begin{bmatrix} e_1 \\ e_2 \\ e_3 \\ e_4 \\ e_5 \\ e_6 \end{bmatrix}
$$

Figure 3.4   Member stiffness array for beam pinned at ''P'' end.

3.2.2 Force–Deformation (Stress–Strain)

The force–deformation equations for an element are

$$\{F\}_{6\times1} = [S]_{6\times6}\{e\}_{6\times1}$$

relating the internal member forces in local coordinates, $\{F\}$, to the displacements of the member in local coordinates, $\{e\}$. Values for the elements of this array are presented in Figure 3.3. The stiffness array for a member that has its P end pinned is presented in Figure 3.4.

3.2.3 Deformation Compatibility

The deformation compatibility equations for an element are

$$\{e\}_{1\times6} = [B]_{6\times6}\{X\}_{6\times1}$$

relating the internal member displacements in local coordinates, $\{e\}$, to the displacements of the system in global coordinates, $\{X\}$.

The transpose of a matrix is obtained by interchanging the rows with the columns. The transpose of $[A]$ is denoted as $[A]^T$. The transpose of a symmetric matrix is equal to the matrix itself. Also,

$$([A] \times [B])^T = [B]^T \times [A]^T$$

The matrix array $[B]$ is the transpose of the equilibrium array $[A]$ or $[B] = [A]^T$. The relation between the equilibrium matrix $[A]$ and the compatibility matrix $[B]$ of the displacement methods can be shown by the law of virtual displacements equating internal and external work to be $[A] = [B]^T$, or conversely, $[B] = [A]^T$. This is an application of Betti's law or its corollary, Maxwell's law of reciprocal deflections.

These relations are combined to form the element stiffness matrix relating the applied loads, $\{P\}$, to the displacements, $\{X\}$, for each member element as follows:

$$\{P\}_{6\times1} = [B]_{6\times6}^T [S]_{6\times6} [B]_{6\times6} \{X\}_{6\times1}$$
$$\{P\}_{6\times1} = [K]_{6\times6} \{X\}_{6\times1}$$

The stiffness array $[K]$ for an individual element can be derived using any classical structural analysis method such as moment–area theorems, Castigliano's theorems, and conjugate beam, slope–deflection, or moment distribution procedures.

3.3 Creating the Global Stiffness Matrix by Assembling Element Stiffnesses

The system stiffness array, $[SS]$, is formed by appropriately placing the components or terms of the $[K]$ stiffness array of each of the members framing into a joint in the global stiffness array $[SS]$ for the system. This stiffness array is denoted as the global stiffness array relating the applied joint loads $\{P\}$ to the joint displacements $\{X\}$ (in the global coordinate system):

$$\{P\}_{NODOF\times1} = [SS]_{NODOF\times NODOF} \{X\}_{NODOF\times1}$$

The $[SS]$ array is always square and symmetrical. A square array is one in which the number of rows equals the number of columns. A matrix is symmetric if terms opposite the diagonal are equal or $a_{ij} = a_{ji}$. Until support conditions are introduced to this global stiffness array, it is singular.

3.4 Solving Simultaneous Equations for Displacements

The system stiffness relation is then solved for the displacements of the system:

$$\{X\} = [SS]^{-1}\{P\}$$

Matrix division is in essence multiplication of an array by an inverse of a matrix. In order for a matrix to have an inverse it must be square and have some properties like positive definiteness that can be determined by the determinant of the matrix.

Cramer's rule is often used to find the inverse of a matrix. Solutions of simultaneous equations can be directly solved for the $\{X\}$ matrix by using Gauss elimination with back substitution. An advantage of matrix inversion compared to Gauss elimination is that many loading conditions or $\{P\}$ columns can later be solved for the internal forces with only one matrix inversion. Likewise for load combinations: when using Gauss elimination, only one pass through the equations is required to solve all the load cases.

Due to its symmetric nature only half of the elements (terms) of the $[SS]$ array need be stored. Within this symmetric half of the matrix there exist many terms whose value is zero. Only the non-zero values need be stored. A "band" of non-zero values is stored, and a banded equation solution routine is used for the symmetric matrix.

3.5 Element Displacements and Forces

The individual element (member) displacements are then

$$\{e\}_{1\times6} = [B]_{6\times6}\{X\}_{6\times1}$$

using the appropriate $\{X\}$ displacements of the joints of the element.

The individual element forces are then

$$\{F\}_{6\times1} = [S]_{6\times6}\{e\}_{6\times1}$$

or directly

$$\{F\}_{6\times1} = [S]_{6\times6}[B]_{6\times6}\{X\}_{6\times1}$$

The definition of each array is given in the program **PFrame**:

$$BT = [B]^T$$

$$S = [S] \text{ and } [B]^T[S][B]$$

$$SB = [S][B]$$

Structural analysis procedures have been denoted as the force and displacement methods. Flexibility coefficients, relating displacement to load, are associated with force methods which determine the system forces and reactions directly. Stiffness coefficients, relating load to displacements, for which displacements are computed first, followed by internal member forces, are the byword for the displacement method. Only the displacement or stiffness method is discussed in this book.

The analysis of beams, continuous beams, and frames, with or without sidesway, follows similar procedures. Joints (nodes) are only required at member slope changes and at support points. Joints can be introduced anywhere within a collinear structure to break it into a series of beams. Since the displacement state described by a beam member (element) is exact, no further accuracy is gained by dividing a structure into smaller beam elements between joints. The only reason to divide a linear beam element into smaller sections would be to obtain the intermediate values for deflection, moment, or shear at these locations.

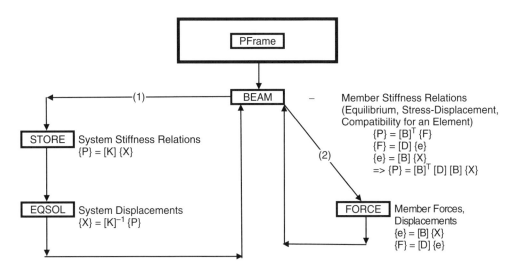

Figure 3.5   Flowchart of VBA program **PFrame**.

3.6 Flowchart of Steps

Figure 3.5 is a flowchart which illustrates the steps required to implement the above-described theory in the VBA program **PFrame**.

 Related Workbook on DVD

The Excel workbook **Load Distribution to Rebar-Concrete** provides a solution that illustrates the concepts presented for the displacement method. The structure consists of a reinforced concrete column under a compressive load P. The portion of the load carried by the concrete, as well as the portion carried by the rebar and the stresses in the concrete and steel, are determined.

References

Buchanan, G.R. (1995) *Finite Element Analysis*, Schaum's Outline Series, McGraw-Hill.

Cook, R.D., Malkus, D.S., and Plesha, M.E. (1989) *Concepts and Applications of Finite Element Analysis*, 3rd edn, John Wiley & Sons, Ltd.

Norris, C.H. and Wilbur, J.B. (1960) *Elementary Structural Analysis*, McGraw-Hill.

Przemieniecki, J.S. (1968) *Theory of Matrix Structural Analysis*, McGraw-Hill.

Zienkiewicz, O.C. (1971) *The Finite Element Method in Engineering Science*, McGraw-Hill, London.

4

Finite Element Analysis VBA Program PFrame

4.1 Program PFrame – Finite Element Analysis (FEA) of Beam–Bar Structural Systems

PFrame is a finite element displacement method program, developed by the author, consisting of procedures written in VBA version 7.0.1590 that is associated with Excel 2010. The VBA program formulates and solves the static equilibrium, force–deformation, and compatibility equations for any two-dimensional (plane) linear structural system composed of beam members. The beam members may have none, one, or both ends pinned. Using this one member type, systems such as determinate and indeterminate continuous beams, frames, as in multi-story buildings or those having non-orthogonal members, and trusses, or combinations of them, may be analyzed.

Internally pinned members are readily accommodated by selecting appropriate member types. Members having variable sections may be analyzed by breaking up a member into a number of different elements. The program can readily solve the problem of imposed displacements or support displacements. Initial pre-stress forces, whether intentional or due to fabrication error, may be imposed.

Data input to the program is from the cells of a worksheet where the input data is entered. Any consistent units that are input will then be output. Input consists of the following variables: joint coordinates, member connectivities, material properties, joint support directions, self-load factor, and the magnitude and direction of loads.

The loads are applied as:

- joint forces;
- member loads consisting of distributed trapezoidal loads over entire member spans or up to two point loads; or
- member fixed-end forces (FEFs).

Solutions for Soil and Structural Systems using Excel and VBA Programs, First Edition. Robert L. Sogge.
© 2012 John Wiley & Sons, Ltd. Published 2012 by John Wiley & Sons, Ltd.

Fractions of member self-loads can be input, if desired. The cells in the area to the right and outside of the input data cells can be used as a scratch sheet for calculations of any kind, as they are ignored on input.

Output is written to cells of the same worksheet used for data input. The structural configuration is plotted on the screen. Output consists of joint displacements (rotations), member forces (moments, shears, axial forces, and axial stresses) at their ends, and support reactions. The total structure weight is computed.

4.2 Creating an Input Data Worksheet

Data input for program **PFrame** is from the cells of a worksheet. Using input from cells of a worksheet in a workbook, rather than from a created data file, is easy to implement and it keeps the code independent of the input data used. Data input should begin where specified in the row and column of the worksheet. This location leaves room for the output plot.

The data worksheet should be named "**PFrame**" on its bottom tab so that plotting can be done directly to that named worksheet. The Excel workbook should be saved with a name describing the input data it contains, using one workbook for each example problem data file. Another workbook named **PFrame**, to which the "Start button" in the data worksheet is assigned, contains the macro program that **PFrame** attached as a VBA module. Only workbooks that contain a macro must be saved as a Macro-Enabled workbook. Those that use or call a macro in another workbook through the assigned Start button need not be Macro-Enabled.

Input data headings are specified so the worksheet can act like an interactive data input creator. The columns adjoining the data can be used to make calculations for the input data values. The output data, including headings, is generated by the macro program and sent to cells of the same worksheet where the input data resides.

4.3 Input Data

The row in the worksheet from which the data input begins is controlled by reading the value in the cell [C8] or =cells(8,3) generated from the statement =row(). The output to the worksheet always begins 11 lines after the last printed line of the input data in the worksheet.

This program coding approach permits a variation in the first data input and output row and the macro program to recognize that difference. The purpose of this construction is to keep the macro independent of the worksheet so that it does not have to be edited for each problem.

Input Data

Title

   1 cell in 1 row

   Title = Problem description title

   The Title parameter may be input with internal commas or quotes.

No. of: (a) Jts, (b) Members, (c) Matls, (d) Suptd Jts

   4 cell values/row

   Jts = number of Joints or nodes

   Members = number of Members or elements having a collinear configuration

   Matls = number of different Material Types

   Suptd Jts = number of Joints which are Supported in any direction.

Joint, X Coordinate, Y Coordinate

   3 cell values/row, as many rows as there are joints

   Joint = number designating Joint

   X coordinate = X coordinate of joint

   Y coordinate = Y coordinate of joint.

   Positive coordinate system is as shown in Figure 3.1.

   Units input = Units output. All input units must be consistent.

   A joint can be placed anywhere within a collinear member.

   Joint Coordinates, Member Numbers, Material Types, and Supports need not be numbered sequentially (in order).

   Any Joint, Member, or Material Type Number cannot be greater than the total number specified.

   Numbering of the joints should be performed in a manner that minimizes the separation between the joints on any one member (element).

Member, P-Jt, Q-Jt, Matl Type, Member Type

   5 cell values/row, as many rows as there are members

   Member = number designating Member number

   P-Jt = number of P Joint for member end

   Q-Jt = number of Q Joint for member end

   Matl Type = Material Type of the member

   Member Type

   = 1 for Beam Member

   = 2 for Beam Member pinned on P end (P end must be denoted as the pinned end)

   = 3 for Beam Member pinned on both ends.

Matl Type, Elastic Mod, Area, Mom Inrta, Wt/Unit L

   5 values/row, as many rows as there are Matl Types

   Matl Type = number designating Material

   Elastic Mod = linear Elastic Modulus of material

   Area = cross-sectional Area of member composed of this material

   Mom Inrta = Moment of Inertia about global Z axis of member composed of this material

   Wt/Unit L = Weight/Unit Length of material.

   As noted previously, Input Units = Output Units. All input units must be consistent.

4.3.1 Member Axis Orientation and Conversion of Moment of Inertia

The moment of inertia for all bending members must be input with respect to the Z global axis with the member axis being aligned along the X–Y axis. If the principal axis of the member corresponds to this Z-axis orientation then no conversion is necessary. Conversions of the moment of inertia from other orientations of a local member coordinate system to the global axis orientation are presented in Chapter 5.

Support Directions (Indicated by 1) for Joint, Horz, Vert, Rotation. Members framing into an external pin support do not need to be specified with pinned-end member stiffness since it is not internal to the structure.

    4 values/row, as many rows as there are supported joints

    Joint = Joint number which is supported

    Horz = 0 if not supported

    = 1 if supported

    Vert = 0 if not supported

    = 1 if supported

    Rotation = 0 if not supported

    = 1 if supported.

Self Load Factor

    1 value

    Self Load Factor = the Factor times the computed member Self Loads that should be used for the load case combination number.

No. of Loaded: Joints, Members, FEF Members

    3 values on 1 line

    Loaded Jts = number of Joints which have Loads (either force P_x, P_y, or moment, M) applied

    Loaded Members = number of Members Loaded with concentrated or distributed loads

    Loaded FEF = number of Members loaded with FEFs.

The following two rows of headings in each of the three applied load categories must be present and separated by a minimum of one row, even if no loads of the specific category are present.

Applied Joint Loads

    Joint, Horz Load, Vert Load, Moment

    4 values/row, as many rows as loaded joints

    Joint = Joint number which is loaded

    Horz Load = magnitude of Horizontal Load applied to joint

    Vert Load = magnitude of Vertical Load applied to joint

    Moment = magnitude of Moment applied to joint.

    The sign convention is the same as for the global coordinate system and is shown in Figure 3.1.

    To input a support displacement, specify the joint as both supported and loaded. The magnitude and direction of the support displacement are input as a load.

Applied Distributed and Point Member Loads
   Member, Dstrb Ld-P, Dstrb Ld-Q, Pt Load-1, Dist-1, Pt Load-2, Dist-2
   7 values/row, as many rows as there are loaded members
   Member = Member number
   DstrbLd-P = Distributed Load at P end of member
   DstrbLd-Q = Distributed Load at Q end of member
   Pt Load-1 = Point Load 1
   Dist-1 = Distance of point load 1 from P joint
   Pt Load-2 = Point Load 2
   Dist-2 = Distance of point load 2 from P joint.
   Member loads (forces) are applied according to the sign convention shown in Figure 4.1.

Applied FEFs
   Member, Moment-P, Moment-Q, Shear-P, Shear-Q, Axial-P, Axial-Q
   7 values/row, as many rows as there are FEF members
   Member = Member number
   Moment-P = fixed-end Moment at P end of member
   Moment-Q = fixed-end Moment at Q end of member
   Shear-P = fixed-end Shear at P end of member
   Shear-Q = fixed-end Shear at Q end of member
   Axial-P = fixed-end Axial force at P end of member
   Axial-Q = fixed-end Axial force at Q end of member.
   FEFs are applied on the joints of the system according to the member sign convention
   shown in Figure 4.2–4.4.

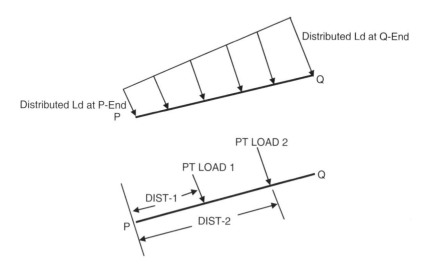

Distributed Ld at Q-End

Q

Distributed Ld at P-End

P

PT LOAD 2

PT LOAD 1

Q

DIST-1

P

DIST-2

Note: Loads oriented to P, Q ends not Global Axis

Figure 4.1   Positive sign convention for applied member loads.

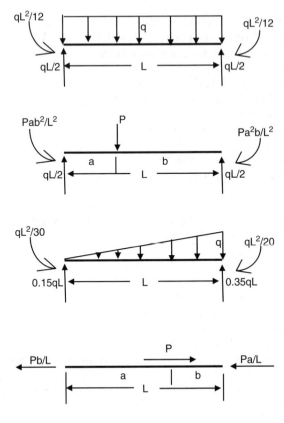

Figure 4.2   FEFs for beam.

4.4 Joint Numbering and Dimensions

In order to minimize computer storage requirements on the SS variable, numbering of the joints should be performed in a manner that minimizes the separation between the joints on any one member (element). Three degrees of freedom are specified at each joint regardless of whether a beam, pinned-end beam, or doubly pinned-end member frames into the joint. The number of simultaneous equations equals 3 × (No. of joints). The bandwidth of the simultaneous equations is based on the maximum joint separation on any one member. Storage requirements for the SS variable equal (No. of equations) × Bandwidth. The structural system stiffness matrix, SS, that needs to be solved is shown in Figure 4.5.

Subscripted variables are automatically dimensioned internally by the program using the ReDim statement. The ReDim statement is similar to the older concept of variable dimensions. As the program is uncompiled the dimensions of subscripted variables can be specified after the input data is read in. This approach automatically and optimally allocates storage size for variables based on what is required for the specific system being analyzed.

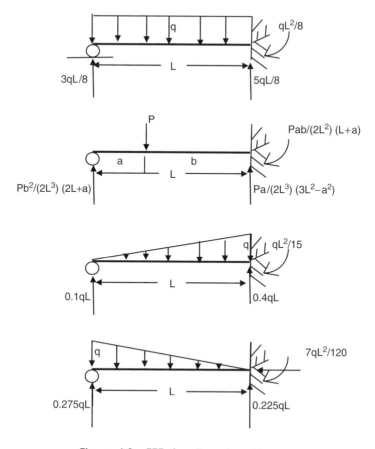

Figure 4.3 FEFs for pinned-end beam.

The number of joints, members, materials, supported joints, or loads does not occupy a lot of computer memory. It is the storage of large arrays that consumes the largest part of computer memory. The VBA program **PFrame** has only one array, SS, that needs to be dimensioned according to the following equations to accommodate a large number of joints (nodes) in the structure:

$$\text{NODOF} = 3 \times \text{NNODES}$$

$$\text{NBW} = 3 \times (\text{NODSEP} + 1)$$

$$\text{NONB} = \text{NODOF} \times \text{NBW} = 9 \times \text{NNODES} \times (\text{NODSEP} + 1)$$

The limit on size appears to be SS(70,000,000) when SS is Dim as a double-precision 8-byte variable. This 0.5 GB Dim size appears to be more related to the Available Physical Memory specified in the Performance tab window of Windows Task Manager.

The Excel workbook **PFrameMemSpace** mentioned in Chapter 2 gives a list of variables in **PFrame** and their storage requirements. If the bandwidth, NBW, is taken as half the number of degrees of freedom or equations, NODOF, then more than 3000 joints can be in the finite element configuration. This number is much more than what would ever be analyzed using a non-graphical data creation worksheet such as the one used for **PFrame**.

A single column matrix can take up less storage space than a square or rectangular array of the same number of elements. Therefore, it is beneficial to store the elements of a square or rectangular array in a single column form and use a conversion routine for converting the elements between the two matrix form configurations. The execution of this approach is shown in program **PFrame** both before and during the EQSOL macro procedure.

4.5 Load Application

Loads can be applied only as joint forces acting directly on the joints in the three global directions.

Loads applied to joints that are supported should not be input since they go directly into the support.

4.5.1 Applied Joint Loads

If a load is applied directly to a joint, and it is not an applied loading resulting from loads intermediate to the joints of a structure, no other forces such as FEFs should be input. For such a loading member forces are computed directly from the displacements resulting from the applied joint loads. This approach is presented in **Example 4.2a**.

4.5.2 Applied Member Loads

All applied member loads must be resolved into forces perpendicular to the member and axial forces along the member. The sign convention for member load application is presented in Figure 4.1. The program will automatically generate the fixed-end moments and shears for these members. The axial forces for inclined members or inclined loads must be computed by the user and input simultaneously either as member FEFs or joint loads. The same example problem loaded by this approach is presented in **Example 4.1** and **Example 4.2b**. Obviously for the loading applied to the model in this section, this approach taken using applied member loads is the most appropriate.

4.5.3 Applied FEFs

When a member is loaded between its end joints the use of a FEF approach to loading must be taken. The theory behind this equivalent loading is derived from the superposition of two force states, one of which has no deformations. A description of this superposition for a simple case is shown in Figure 4.6.

Example 4.1 shows the result when this superposition is done automatically by program **PFrame**. Input and output for **Example 4.1** are shown in the Excel worksheets.

The Excel data file should be created as indicated. Data input for all the examples is supplied on the program disk. Reference to these problems provides a simple self-learning approach to understanding the program.

 Example 4.1   Beam on pinned supports.

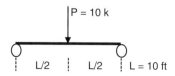

FEFs are required for both ends of any member, whether supported or not. This approach of applying loads is presented in **Example 4.2c**. This example is for instructional purposes only. Obviously for the loading applied to this example, the approach taken using applied member loads in the previous section is the most appropriate.

 Example 4.2   Beam 3 loading approaches (a) Loaded Joint Approach (b) Loaded Member Approach (c) Fixed-End Force Approach.

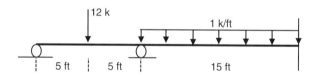

If the unbalanced FEFs resulting from loads applied between member joints are applied directly to the joints, it is necessary to add the FEFs to the computed member forces to obtain the final member force. For such cases the FEF state is input by the user. The program automatically loads the joint and adds the member FEFs to the member forces for an applied joint load to yield the final member force.

The FEF states for four general loading conditions, from which others can be superimposed, are presented in Figures 4.3–4.5. Figure 4.3 is for a beam with moments on its ends, Figure 4.4 is for a member pinned at one end, and Figure 4.5 is for a member pinned at both ends.

4.6 Imposed Joint Displacements

A joint displacement can be imposed by specifying the coordinate direction of the displacement as a support coordinate direction and by inputting the magnitude of the

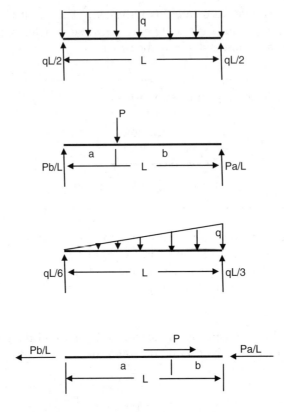

Figure 4.4 FEFs for pinned–pinned beam.

displacement as a load on the joint in the same coordinate direction as the support displacement coordinate direction. An example of an imposed support displacement is presented in Chapter 7.

4.7 Unstable or Improperly Supported Configurations

The program checks to see that stiffness of the system provides a value on the diagonal of the stiffness matrix. If zero exists on the diagonal, infinite displacements will occur due to the system being supported improperly or in an unstable configuration.

4.8 Running Program PFrame

The macro VBA program can be initiated from the data worksheet using the Start button. The Start button must be associated with the **PFrame** macro. The easiest way to

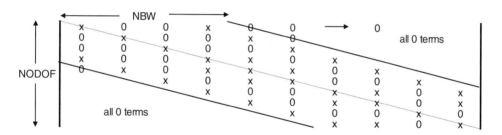

Figure 4.5 Stiffness Matrix Storage.

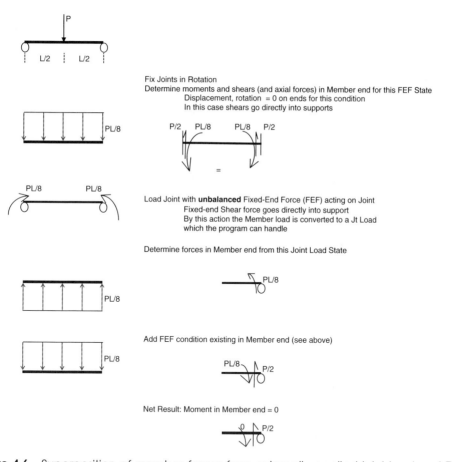

Fix Joints in Rotation
Determine moments and shears (and axial forces) in Member end for this FEF State
        Displacement, rotation = 0 on ends for this condition
        In this case shears go directly into supports

Load Joint with **unbalanced** Fixed-End Force (FEF) acting on Joint
        Fixed-end Shear force goes directly into support
        By this action the Member load is converted to a Jt Load
        which the program can handle

Determine forces in Member end from this Joint Load State

Add FEF condition existing in Member end (see above)

Net Result: Moment in Member end = 0

Figure 4.6 Superposition of member forces from externally applied joint load and FEF states.

properly assign the Start button is to open the program containing the VBA macro sub **PFrame**. Then by right clicking on the Start button in the worksheet containing the input data the macro VBA program **PFrame** can be initiated. Execution of the macro program is initiated by clicking on the Start button of the data worksheet. This action causes the workbook containing the required macro program **PFrame** to be automatically opened, and the macro executed. Once the Start button is assigned it is not necessary for the workbook containing **PFrame** to be open in order to run the macro.

Viruses have often targeted the macros in Excel files. To permit the opening of macros within Excel it is necessary to adjust its security levels to a lower level acceptable for opening macros. Adjust the security level in Excel using

```
Tools > Macro > Security > medium or low - Excel 2003
Developer > Macro Security > Trust access to the VBA project
object model - Excel 2010
```

4.9 Output Data

The first output consists of a plot of the structure to the worksheet from sub PLOT:
Plot of structure – solid line for beams, dashed line for bars.
The structure is plotted on the data input worksheet named **PFrame** using the AddLine feature of VBA. The plot is in an area 10 in wide by 6 in high on the screen. The actual size that is plotted depends on the points/inch plotted on the screen.
The program is based on:

Horizontally – 100 points = 133 plotted points = 2.09in $\Rightarrow$ 47.8 points/in on screen.
Vertically – 100 points = 133 plotted points = 2.09in $\Rightarrow$ 57.1 points/in on screen.

The plot is not centered in the box but is drawn beginning at the origin at (1,3) inches on the screen or approximately column B and row 14.
The program plot width is based on

```
myDocument.Shapes.AddLine(XP(IPP), YP(IPP), XP(IQQ), YP(IQQ)).Line.
DashStyle = msoLineSolid.
```

The name in the following is the name on the tab of the worksheet getting the output plot:

```
Sheets("PFrame").Select
Set myDocument = Worksheets(1)
```

The output is written to the worksheet cells beginning at a row 10 rows below the last input line. Since this output data is associated with some charts of displacements, and member forces, the location within the worksheet of the cells with the source data that is to be plotted must be at the location associated with the chart.

Output Data

No. of Eqns, Bandwidth, Jt Sep
   No. of Eqns = 3 × No. of Jts
   Bandwidth = 3× (Maximum Joint Separation on any one member) − 2
   Jt Sep = Joint Separation on any one member.
Joint Displacements in Global Coordinates (Figure 3.1)
   Units Output = Units Input Rotations are in radians.
Member Forces – Moments, Shears, Axial Force, and Stress in local coordinates (Figure 3.1)
   Note that the local coordinate system for a member is a function of the P and Q joint
   orientation of the ends of the member.
Support Reactions in Global Coordinates (Figure 3.1)
   The magnitude of the reaction force at each support is computed from the member forces.
   The total reaction is computed by adding that part of the FEF which would have been
   transferred to the support.
Weight of Structure
   The total weight of the structure is computed using the weight per unit length for each
   material type.

4.10 Alternate Solution Approach to Macro Program PFrame

The setup of the constitutive equations that need to be solved can be performed exclusively within the worksheet without using any associated macro like the VBA program **PFrame**. Since Excel has the ability to invert a matrix and multiply matrices using the math functions within the worksheet, the deformations along the pier can be developed internally within the worksheet and an external equation solver is not needed. Such a computation would apply to a system configuration for which a generalized solution platform can be erected. For such a system the constitutive equations are set up and then solved for displacements using the INVERSE function available in Excel. The internal member forces are computed from the displacements using the MMULT math function of Excel.

Examples that show the use of the MINVERSE and MMULT functions available in Excel are Examples 6.1a, Example 7.1a, Example 16.2b using a beam-on-elastic foundation model and in Section 18.11 with Example 18.7b for a laterally loaded pile. These examples show two different ways of solving a problem; the usual way using program **PFrame** and using the MINVERSE and MMULT functions available in Excel. The latter approach sets up the constitutive equations and solves them directly using the Excel functions.

4.11 Significant Aspects of Excel Worksheet & VBA Macro Program Construction

This section summarizes the specific program code constructions that have been employed in the development of Program **PFrame** and that are used in this part of the text in the analysis of structures.

- Shows use of a VBA macro program with input totally from cells of a worksheet. It is best to create data for input to a macro program directly in a worksheet since headings for data can be given and the worksheet acts like an interactive data file input creator. The problem output data drive\location and filename are input from worksheet cells to the macro. This approach allows the program to stay independent of having to be edited. Such input is easier than using an Input Box and is consistent with input from the rest of the worksheet. On output, rather than create a sequential data file it is easiest to just output the results directly to the cells of the worksheet. This approach requires that the problem input data be saved with a workbook name, one for each problem data file.
- The line in the worksheet from which the data input begins is input to the macro by reading the value in the cell generated from the statement =ROW(). This construction allows for a variation in the first data input lines and the macro program to recognize that difference. It also eliminates any need to edit the macro code lines, keeping the macro independent of the worksheet.
- The macro program is started with a "Start" Button assigned to the **Pframe** macro.
- Plotting structure in Worksheet using line commands in macro SUB PLOT.

```
With myDocument.Shapes.AddLine(XP(IPP), YP(IPP), XP(IQQ),
   YP(IQQ)).Line
.DashStyle = msoLineSolid
The name in the following name is the name on the tab of the
   worksheet getting the output plot
Sheets("Arch").Select
Set myDocument = Worksheets(1)
Solid lines are plotted for beams, dashed lines for bars.
```

- Use of Sub Procedures in Program **Pframe**. Their difference from a Function Procedure will be shown in Chapter 24. In that chapter a good example of a Function Sub Procedure is in the iterative routine used to calculate interest in program Loan. At this point the following construction shows a Function Procedure:

```
Function C(A as Single, B as Single), as Single
C = SQRT(A2 + B2)
End Function
```

In the worksheet, = Function C() in a cell, returns value to that cell
Although they look similar to a Sub Procedure, they are different in that a Sub Procedure will not return values to a cell.
- Simultaneous equation solution using Gauss elimination with back substitution.
- Multiplication and Inversion of matrices using the math functions MMULT and MINVERSE.
- Use of Double versus Single precision.

5

Beams

This chapter will go into a little more detail about the application of the various types of beams.

5.1 Beam Member Types

The following three types of rotational end restraints on beam members may be specified by the member type (NTYPE) parameter:

Member Type Rotational End Restraint

1. Beam – moment on both ends
2. Beam – pinned at P end
3. Beam – pinned at both ends.

For members with only one end pinned, the P end must be denoted as the pinned end.

It is not necessary to define a member as a pinned-end beam unless that pin is on the end of a member internal to the frame system. Where an external pinned support is supplied, specifying the horizontal and vertical supports and the rotation as unsupported is adequate to properly develop the stiffness of the system. In such cases the rotation at the externally supported joint will be automatically computed.

Fixed-end forces (FEFs) for beams and those with singly and doubly pinned ends have been presented in Figures 4.3–4.5.

5.2 Bar Members as Pinned-End Beams

A bar member is a beam member with a moment of inertia equal to zero that has its ends supported against rotation. Alternatively, instead of providing rotational joint supports a small moment of inertia value could be specified.

Solutions for Soil and Structural Systems using Excel and VBA Programs, First Edition. Robert L. Sogge.
© 2012 John Wiley & Sons, Ltd. Published 2012 by John Wiley & Sons, Ltd.

A bar element can also be created by using a Member Type = 3 element (beam with both its ends pinned) with the member's moment of inertia set equal to zero. Member moments of inertia (Mom Inrta) can equal zero as long as any joints that such members frame into are supported or have at least one member with a rotational resistance framing into it. When only pinned-end members frame into a joint, the rotational degree of freedom of the joint must be supported.

The program checks to see that stiffness of the system provides a value on the diagonal of the stiffness matrix. If zero exists on the diagonal, infinite displacements will occur due to the system being supported improperly or being in an unstable configuration.

Examples 5.2 and **5.3** show applications to structures with pinned ends.

Example 5.1   Beam as one or two-member structure (a) Load Input by Loaded Member of FEF Member (b) Load Input by Loaded Joint.

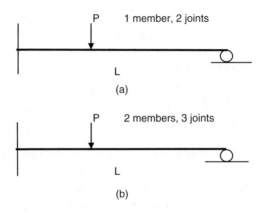

(a)

(b)

Example 5.2   Beam with internal hinge (PFIbm).

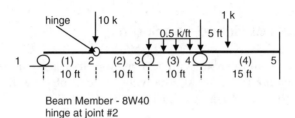

Beam Member - 8W40
hinge at joint #2

Example 5.3   Doubly pinned-end beam (a) on column supports (b) on supports.

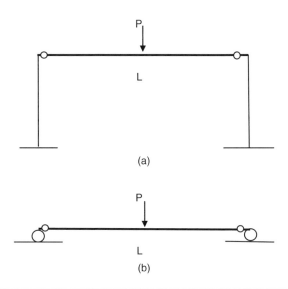

5.3 Moment of Inertia Conversion for Different Member Axis Orientation

The moment of inertia for all bending members must be input with respect to the Z global axis. If the principal axis of the member corresponds to this orientation then no conversion is necessary. Conversions from other orientations of a local member

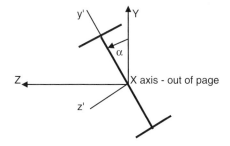

$$I_{zz} = \sin^2 \alpha\, I_{y'} + \cos^2 \alpha\, I_{z'} - 2\, I_{yz} \sin \alpha \cos \alpha$$

Figure 5.1   Moment of inertia conversion for different member axis orientation.

coordinate system to the global axis orientation can be made using the following equation presented in Figure 5.1:

$$I_{ZZ} = \sin^2 \alpha I_{y'} + \cos^2 \alpha I_{z'} - 2I_{yz} \sin \alpha \cos \alpha$$

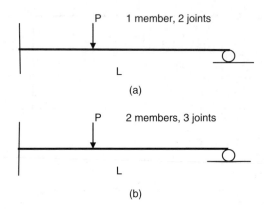

Figure 5.2   Load application for one- and two-member systems.

Example 5.4   Cantilevered beam with distributed load.

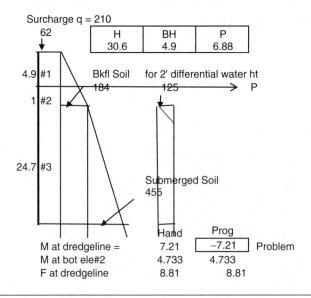

5.4 Load Application

If a load is applied directly to a joint and is not an applied loading resulting from a fixed-end force state then no fixed-end forces should be input. No internal superposition of fixed-end forces and computed member forces is required for such applied joint loads. An example of this situation is presented in Figure 5.2 and in **Example 5.1** where the same problem is approached using a load application by the applied member load method on one member and by using an applied joint load on the structure divided into two members to achieve the same result. Another example of a distributed load application is presented in **Example 5.4.**

 Related Workbooks on DVD

Example 5.1a – Beam 1 mbr 2 jts
Example 5.1b – Beam 2 mbrs 3 jts
Example 5.2 – PFIbm
Example 5.3a – Pinned Beam on Supts
Example 5.3b – Pinned Beam on Column Supts
Example 5.4 – Cantilevered Beam with Dist Ld

6

Frames

6.1 Analysis of Frames

The analysis of frames, with or without sidesway, uses the same procedures given previously for beams. Since the displacement state described by a beam member (element) is exact, no further accuracy is gained by dividing a structure into smaller beam element divisions between joints. Joints (nodes) need only be inserted at member slope changes and at support points.

As with a beam, each joint is specified as having three degrees of freedom: the horizontal lateral motion (often referred to sidesway for a frame), a vertical deflection, and a rotational deflection. Thus the sidesway occurring during the loading of any determinate or indeterminate frame system will be automatically computed.

6.2 Rigid Joints

The rotational deflection of the joint assumes that the joint is rigid – that the ends of all the members framing into the joint rotate equally. Similarly, a frame is defined as rigid when all joints within the frame have rigid connections. To simulate or allow some rotation of a member at a joint of the structure, a bar member having one pinned end is used. Number the joints on the member so the pinned end (P joint) frames into the joint.

6.3 Joint Numbering

With frame structures, numbering the joints so as to minimize the separation between the joints on any one member becomes critical compared to the numbering for a beam alone. This topic was introduced in Section 4.4. An example of an efficient joint numbering pattern for a frame is shown in Figure 6.1.

Solutions for Soil and Structural Systems using Excel and VBA Programs, First Edition. Robert L. Sogge.
© 2012 John Wiley & Sons, Ltd. Published 2012 by John Wiley & Sons, Ltd.

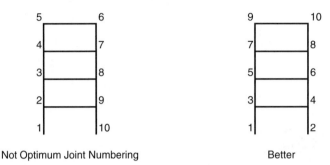

Figure 6.1   Optimum joint numbering for minimum node separation and bandwidth.

The number of simultaneous equations equals the number of degrees of freedom (NODOF):

$$\text{NODOF} = 3 \times \text{NNODES}$$

The bandwidth of the simultaneous equations, NBW, is based on the maximum joint separation (NODSEP) on any one member:

$$\text{NBW} = 3 \times (\text{NODSEP} + 1)$$

The size of the SS matrix is then

$$\text{NONB} = \text{NODOF} \times \text{NBW} = 9 \times \text{NNODES} \times (\text{NODSEP} + 1)$$

As noted previously, the number of joints, members, materials, supported joints, or loads does not occupy a lot of computer memory. It is the storage of large arrays that consumes the largest part of computer memory. The VBA program **PFrame** has only one array, SS, that needs to be dimensioned to accommodate a large number of joints in the structure. Again as noted previously, the Excel workbook **PFrameMemSpcCalc** gives a list of variables in PFrame and their storage requirements.

If the bandwidth, NBW, is taken as half of the number of degrees of freedom or equations, NODOF, then greater than 3000 joints can be in the finite element configuration. This number is much more than what would ever be analyzed using a non-graphical data creation worksheet such as the one used for PFrame. The number of members, if taken as approximately equal to the number of joints, is also very large.

Since the number of joints and thus members which can be analyzed is rarely a limitation, procedures such as the following are not required:

- storing joint and member information on external files and calling it when needed;
- using an out-of-core equation-solving routine that solves a portion of the equations and stores them in a file;
- ignoring the compression of an individual member so that the degrees of freedom of a system can be reduced by a third;
- using a bandwidth minimizer that automatically numbers the joints in an optimal manner to allows more variable storage space within the computer memory core for the simultaneous equations.

6.4 Pinned-End Beam

Again, as is the case with continuous beams, the following three types of rotational end restraints on members may be specified by the member type (NTYPE) parameter:

Member Type Rotational End Restraint

    1. Beam – moment on both ends
    2. Beam – pinned at P end
    3. Beam – pinned at both ends.

Regarding external pinned support points on a frame, in order to simulate a beam connected at one end by a pin or roller support it is best to use a regular beam (Member Type $= 1$) and specify the rotational direction as unsupported. For this support condition the rotation at the pinned end of the member will be determined by the program. If a pinned-end beam were used to model this situation then the rotation would not be determined. When pinned-end members only frame into a joint, the rotational degree of freedom of the joint must be supported. Examples of frames that require the use of a pinned-end beam are presented in Figure 6.2.

The two system structural conditions of internal pins shown require that a fixed-pinned element stiffness be used for members framing into the pinned joint

Figure 6.2   Frame structures requiring pinned-end member.

6.5 Supports

6.5.1 Inclined or Skewed

Supports that do not restrict motion only in the horizontal or vertical direction are referred to as inclined or skewed supports. In order to input inclined supports, an engineering simulation approach is best. A real-life engineering simulation for two inclined support conditions is shown in Figure 6.3. By using a large member area (Area) no displacement in the support direction will occur. The small moment of inertia (Mom Inrta) will cause no moment to develop at the bottom end of the column.

Alternatively, the element stiffness array could be multiplied by coordinate transformation arrays (similar to member conversion to I_{ZZ} in Chapter 5) as follows. Changes to the program would be required for the implementation of coordinate transformation arrays.

$$x' = x \cos \theta + y \sin \theta$$

$$y' = -x \sin \theta + y \cos \theta$$

In matrix notation:

$$\begin{bmatrix} x' \\ y' \end{bmatrix} = \begin{bmatrix} \cos \theta & \sin \theta \\ -\sin \theta & \cos \theta \end{bmatrix} \begin{bmatrix} x \\ y \end{bmatrix}$$

In order to invert the above transformation matrix put in $-\theta$ instead of θ. This works since the matrix is orthogonal.

$$\begin{bmatrix} x \\ y \end{bmatrix} = \begin{bmatrix} \cos \theta & -\sin \theta \\ \sin \theta & \cos \theta \end{bmatrix} \begin{bmatrix} x' \\ y' \end{bmatrix}$$

6.5.2 Elastic

Supports providing some partial support restraint can be described as supports having an elastic property. This elastic property can be in their translational or rotational directions. Systems with such restraint can be analyzed in either of the following two ways.

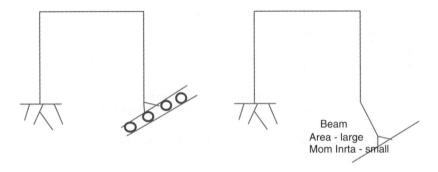

Beam
Area - large
Mom Inrta - small

Figure 6.3   Inclined support representation.

The easiest way to make a model that simulates the elastic support properties is to use beam members connecting to the joint. The beams should have the desired AE/L to simulate the elastic stiffness in the translational direction, and $4EI/L$ to simulate the rotational direction elastic stiffness.

Alternatively, elastic support conditions can be input by modifying the program such that the desired stiffness is added to the element of the system stiffness array which represents the proper coordinate direction. The stiffness should be added to the stiffness matrix in the sub STORE before entering the sub EQSOL. The proper element, a, of the singly subscripted stiffness array corresponding to the coordinate direction, b, is

$$a = \text{NBW} \times (b - 1) + 1$$

where b, the coordinate direction, is defined as follows:

Horizontal: $b = 3 \times$ Joint Number -2
Vertical: $b = 3 \times$ Joint Number -1
Rotational: $b = 3 \times$ Joint Number
NBW = Bandwidth of Stiffness Matrix $= 3 \times$ (Jt Sep $+ 1$).

When modeling a spring restraint on a joint the elastic support stiffness should be added to the stiffness matrix, SS, before entering EQSOL.

6.5.3 Imposed Support Displacements

Beam and frame systems having supports with known displacements are analyzed in the same manner. Support displacements are input into the program in a simple way that decouples the stiffness equations in the support degree of freedom direction. Consider a support displacement point as both a support point and a load point when inputting the number of supports and the number of loads. Then input the support displacement value in the load vector. The program multiplies the load vector and the term on the diagonal of the structure stiffness matrix by a very large number, 10^{10}. This multiplication in effect disengages that equation from the set of stiffness equations, and it yields the imposed deflection as the deflection at the supported joint where the support displacement is imposed.

Alternatively, where support displacements are zero the applicable elements of the stiffness array and load vector array are removed from the arrays by separating the equations from the others. This decoupling is accomplished by putting a one on the diagonal and zeroing out every other element in the same row of the load and stiffness arrays, as well as the same column of the stiffness array.

6.6 Varying *EI* of Members Comprising a Frame

A similar case to elastic supports occurs in any indeterminate frame system having members of different stiffness properties. The proper distribution of member forces in the frame to the applied loading is dependent on the load path taken by those externally applied forces. The load path taken by the forces to the system supports is governed by the relative stiffness of the members of the structure. Thus it is important to define precisely the parameters E, I, and A, on which frame stiffnesses are dependent for each member.

Members that have a varying section property can be represented by breaking the one member into a series of members with different material and section properties.

6.7 Stability – The $P-\Delta$ Effect

Instantaneous loading is only conservative if the initial geometry of the structure and the final deformation configuration are essentially the same, in magnitude as well as shape. The deformation of a frame itself can cause increased forces to exist in members of a frame. Such forces are typically referred to as secondary forces. They can be determined using equilibrium and compatibility arrays based on the deformed joint coordinates. This type of analysis incorporates the $P-\Delta$ effect. After computing the displacements, the member strains and then forces can be computed using a compatibility matrix $[B]$ based on the new deformed joint coordinates. For most cases the displacements are assumed to be small and this correction for secondary forces is not required.

6.8 Load Case Combinations of Load Groups

Although the analysis of the load groups feature has not been implemented in the **PFrame** program in this text, it can be readily programmed and added. Load groups consisting of load case combinations derived from various load cases can be handled efficiently on one pass through the in-core equation solver program. The support directions and locations in each of the load cases must be identical. Input would consist of the number of load case combinations, load cases, and the load case factors.

For each load case combination the loads may be applied simultaneously to joints by specifying (Loaded Joints). To members they may be applied either as fixed-end forces (FEFs), by specifying (FEF Members), or as distributed and concentrated member loads, by specifying (Loaded Members). Load case combinations of applied member loads must not result in more than two point loads at different locations on any member. If a point load on a member in a load case is Pt Load 1, then another point load on the same member in another load case must be denoted as Pt Load 2. If more than two concentrated forces are applied to a member, that member can be broken down into smaller individual member segments. Each segment would contain at most two concentrated loads and a distributed load having a trapezoidal distribution.

Displacements can be imposed on any supported joints within any load case regardless of the number of load cases or load case combinations. The only requirement for imposed displacements is that the coordinate direction of the joint to which the displacement is imposed is a supported coordinate direction in all load cases. If a displacement is imposed on any joint that is not supported in all load cases, then such an analysis must be conducted on a single load case combination consisting of a single load case. Therefore, an imposed lateral displacement at the top of a bent cannot be analyzed except as a single load case combination consisting of a single load case.

6.9 Interior Member Forces

Interior member forces consisting of moments, shears, and displacements can be determined exactly by using the equilibrium equations on the statically determinate segment. The segment has forces that consist of the member end moments and shears and the load applied between the ends of the section being considered.

For member types 1 and 2 the elastic load procedure (Norris and Wilbur, 1960) can be used to compute the approximate slopes and deflections at the quarter points. A linearly varying moment diagram is assumed to exist between the ends of each segment. Slopes (radians) are determined with respect to the original undeformed geometry. For member type 1, deflection is measured as the perpendicular distance of the deformed member shape from a chord extended through the P joint and Q joint in their undeformed position. For member type 2, deflections are measured from a chord extended through the Q joint in a direction parallel to a chord between the undeformed joints. The positive sign convention for interior member forces, slopes, and deformations is shown in Figure 4.2. The sign convention for members places the P(i) joint on the left and the Q(j) joint on the right.

For members pinned at both ends (NTYPE = 3), the displacements at the quarter points are measured from a chord connecting the original member endpoint locations. They are computed using the exact elastic equations. For this type member, no rotations at the quarter points are computed.

Analysis of interior member forces, deformations, and slopes at quarter points along a member has not been implemented in the **PFrame** program in this text. It can be readily programmed and added. Alternatively for this program, interior member forces at various points within the member can readily be obtained by dividing that member into many smaller members using additional joints along the member. This approach does not require significantly more computational or input preparation time and provides additional information with ease.

The deformed shape (deflection), moment, shear, and axial load of the structure are not plotted. Such parameters can readily be graphically displayed using Excel's charting capability. In later chapters such charting is performed for the output from soil–structure interaction examples.

6.10 Examples

Application of the **PFrame** VBA macro program to frame systems is presented in the following examples. The input is from cells in the worksheet named for the example.

The required input is indicated by the headings in the Excel worksheet. Input and output can, but not in this PFrame application, be written to a sequential data file that can readily be printed using Notepad. Reference to these examples provides a simple self-learning approach to understanding the program.

It would be beneficial to store the examples in separate worksheets all contained in one workbook labeled Chapter 4-Examples. This approach is not possible for the reason that the statements:

```
Sheets("PFrame").Select
Set myDocument = Worksheet(1)
```

that are required for the PLOT macro of program PFrame cause a problem with the requirement that one cannot rename worksheets to the same name as another worksheet or a workbook referenced by Visual Basic. It would be necessary to modify these statements in the PLOT routine of PFrame each time removing the independent nature of this macro. Therefore each example is stored in a separate workbook with the worksheet labeled "PFrame".

Examples 6.1a and **b** Fixed Frame take a beam member supported by beams applied at both ends and determine, for the load conditions of lateral and moment joint loads, the displacements at the two unsupported joints and the forces in the beams by using:

a direct stiffness approach that employs the math functions MMULT and MINVERSE that were presented in Chapter 1;
a finite element solution using program PFrame.

Examples 6.2a and **b** take a portal frame having fixed supports and determine the displacements and member forces in the beams by using:

a direct stiffness approach that employs the math functions MMULT and MINVERSE that were presented in Chapter 1;
a finite element solution using program PFrame.

By changing the relative stiffness of the beam in relation to those of the column members in this example problem, the changes that result in the member force mode, whether moment, shear, or axial, can be observed. The resulting changes to the reaction forces are noted and can be plotted as a function of the stiffness ratio, I_{beam}/I_{column}.

Example 6.3 consists of a portal frame with a pinned support. Note that the right column that is supported by an external pin support does not need to be specified with pinned-end member stiffness since it is not internal to the structure.

Example 6.4 deals with a frame having a non-rectangular shape. The solution to this frame example is given in Problem 6.8.2 loading condition 1, page 101 of Wang (1966). This problem presents a more complex example of a FEF loading for a frame. The frame has inclined members with vertical and horizontal point and distributed loads.

This frame can also be analyzed by dividing the top member into two members and applying a joint load at the member juncture.

Example 6.5 covers a bike frame.

In **Example 6.6**, a bridge bent supported by drilled piers, the columns are extended beneath a river bottom level to a distance equal to the point of fixity. (Point of fixity will be discussed in Chapter 18.)

In **Examples 6.7a** and **b**, for an arch simulated by two bars and beams, the arch structure is composed of either two bar or two beam elements. There is no difference if the top connection of the two beam elements is modeled as being rigid or as being pinned. Therefore this arch structure can be treated as a three-hinged arch for the symmetrical loading.

In **Example 6.8**, for an arched frame supported laterally by elastic restraint, the arch frame is represented by elements whose end joint coordinates are derived by dividing chords of the circular arc into 10 linear beam element segments. The loading for the arch consists of a 10 ksf uniform horizontally distributed vertical load that is not attenuated. Due to the symmetry of both the configuration and loading, the top joint is supported horizontally and in rotation. The bottom of the arch is supported laterally by an elastic support consisting of a spring. A beam element, having no moment of inertia (essentially a bar), simulates this spring.

The stiffness of the arch–frame structure is varied by using different parameters for the beam's E, I, and A. The stiffness of the spring is varied by using different parameters for the bar's E and A. These variations are shown in the worksheet. Using this model, M_{top} in the arch and the horizontal force in the spring versus the ratio of the spring stiffness (kA) to the structure stiffness (EI/span^3) are developed and plotted in Figure 20.1.

 Related Workbooks on DVD

Example 6.1a Fixed Frame
Example 6.1b Fixed Frame
Example 6.2a Portal Frame
Example 6.2b Portal Frame
Example 6.3 Frame Pinned Supt
Example 6.4 Frame having Non-rectangular Shape (PFIfram)
Example 6.5 Bike Frame
Example 6.6 Bridge Bent
Example 6.7a Bar Arch
Example 6.7b Beam Arch
Example 6.8 Arched Frame Supported Laterally by Elastic Restraint (ArchElasSupt)

References

Norris, C.H. and Wilbur, J.B. (1960) *Elementary Structural Analysis*, 2nd edn, McGraw-Hill.
Wang, C.K. (1966) *Matrix Methods of Structural Analysis*, International Textbook Company.

7

Trusses

7.1 Theory for Bar Members

Determinate or indeterminate trusses composed of axially loaded bar members can be analyzed using program **PFrame**. The indeterminacy of a truss may be with respect to external reactions or internal bar forces.

PFrame is developed for beam elements having three degrees of freedom (DOF) per joint. Using a program based on beam members having three DOF per joint is not as efficient in storage as a program developed with a stiffness matrix for a bar member having only two DOF per joint. This loss of storage efficiency and potential inability to handle extremely large trusses is rarely a problem.

The analysis of a truss using **PFrame** is conducted by eliminating the rotational DOF of the beams. Implementation of the truss analysis can be conducted in any of the following three ways:

- The beam can be made very flexible by setting its moment of inertia equal to zero. All rotational DOF at all joints must be eliminated by supporting in the rotational direction.
- Alternatively, yet less exact, the moment of inertia can be set to a very small number. With this approach the rotational support of the joint is not necessary.
- Create a bar member by setting (Member Type = 3). With this approach the joint rotational DOF must be supported. The moment of inertia (MOM INRTA) can equal zero.

7.2 Analysis of Bar Assemblage

Example 7.1 takes a simple assemblage of bars and determines the forces in them using:

(a) a direct stiffness approach that employs the math functions MMULT and MINVERSE that were presented in Chapter 1;
(b) a finite element solution using program **PFrame**.

Solutions for Soil and Structural Systems using Excel and VBA Programs, First Edition. Robert L. Sogge.
© 2012 John Wiley & Sons, Ltd. Published 2012 by John Wiley & Sons, Ltd.

7.3 Load Application

With a truss structure non-axial loads cannot be applied between the joints. For the case of an applied axial force, intermediate to the joints, the fixed-end force load application is identical to that shown for a beam member with both ends pinned in Chapter 3. These fixed-end forces for a bar element member are shown in Figure 4.4. **Examples 7.2a** and **7.2aa** are demonstrative of loading a truss at its joints. The examples take two of the approaches discussed in the previous section to represent the member properties, one using type 3 members and another using type 1 members with $I = 0$.

7.4 Initial Member Length Changes

Indeterminate trusses that are constructed using members having initial fabrication elongations or shortenings will induce stress and strains in internal members. These changes in the member lengths could arise from either fabrication errors or intentionally wanting to induce a pre-stress force in the member. Either applied fixed-end axial forces or joint loads can be used to simulate such cases.

By the fixed-end force approach, axial forces would be applied to the member that would strain it to match its initial geometry. Compressive forces would be used for members fabricated too long, tensile forces for members fabricated too short. The total internal force in the member would be computed by the program by summing the computed member force and the applied fixed-end axial force. **Example 7.2c** shows an initially elongated member for a truss by the fixed-end force approach.

Alternatively, joint forces which would pull the member end joints apart for initially elongated members, and push the member end joints together for initially shortened members, could be applied. The resultant internal force is equal to the computed (output) member forces due to the applied joint loads. **Example 7.2c** shows the joint force approach to member elongations.

7.5 Support Displacements

Truss support displacements will only induce internal forces in trusses that are statically indeterminate with respect to reactions. Support displacements can be implemented as specified previously for program **PFrame**. **Example 7.2b** covers the settlement of a support of a truss.

Reference (for Example 7.2)

Wang, C.K. (1966) *Matrix Methods of Structural Analysis*, International Textbook Company, Scranton, PA, p. 56.

8

Reinforced Concrete

8.1 Concrete and Reinforcing Steel Properties

The strength of reinforced concrete can be described by a stress–strain curve that fits a parabolic equation. This curve is shown in worksheet Stress-Strain Prop of the workbook **Reinf Concrete**. The maximum usable extreme concrete compression fiber strain occurring at maximum stress is generally taken as $\varepsilon_c = 0.003$ for all strengths of concrete (Article 10.2.3 ACI, 1989, Article 5.7.2 AASHTO, USACE, 1992). The elastic modulus of concrete, E_c, is customarily defined for the linear secant value at a stress of half of the compressive strength or $0.5f_c'$. For the reinforced concrete section a linear E is computed from the concrete compressive strength, f_c', and its unit weight, w_c (in lb/ft^3), as

$$E_c = w_c^{1.5}\, 33(f_c')^{0.5}$$

With w_c, the unit weight of plain concrete in lb/ft^3, typically equal to 145, this relation becomes

$$E_c = 57,600(f_c')^{0.5}$$

The minor increase in the elastic modulus of a beam element acting in plate action due to plane-strain conditions, of $1/(1 - v^2)$ or approximately 3%, is generally ignored. Other related properties of concrete are:

Poisson's ratio $v = 0.17$
Tensile strength $= 0.10 f_c'$
The modulus of rupture, f_r, is defined as the flexural tensile stress at which concrete first cracks (in bending or flexure mode), thus

$$f_r = (7.5 \text{ to } 10)(f_c')^{0.5} \text{ (psi)}$$
$$f_r = (0.20 \text{ to } 0.37)(f_c')^{0.5} \text{ (ksi)}$$

Solutions for Soil and Structural Systems using Excel and VBA Programs, First Edition. Robert L. Sogge.
© 2012 John Wiley & Sons, Ltd. Published 2012 by John Wiley & Sons, Ltd.

Coefficient of thermal expansion $\alpha = 6 \times 10^{-6}/°F$

Similar properties for the steel reinforcing are:

$E = 29,000$ ksi
$\nu = 0.27$
$f_y = 60$ ksi
$\gamma = 492$ lb/ft^3
$\alpha = 6.5 \times 10^{-6}/°F$

Since the coefficient of thermal expansion, α, is relatively equal for concrete and steel reinforcing, no cracking occurs at the interface of the materials during large changes in temperature.

A balanced design is defined where the steel reaches its yield strength simultaneously with the concrete reaching its compressive strength. The ratio of the reinforcing steel area to the area defined by bd is defined, for a rectangular section at balanced conditions, as

$$\rho_b = 0.85\beta_1(f_c'/f_y)87/(87+f_y)$$

This ratio equals 2.14% for tension reinforcing steel where $f_y = 60$ ksi, and $f_c' = 3$ ksi, for which the factor $\beta_1 = 0.85$ (refer to Section 8.6 Notation for a description of the terms used). Reinforcing quantities as a percentage, p, of A_g, not bd, are given in the workbook **Reinf Concrete**. Table 8.1 presents the minimum concrete cover distance from the outside of the reinforcing bar.

The workbook **Reinf Concrete** contains other various design parameter relations such as steel reinforcing sizes, A_s/ft, lb/cy, lap splice length, and welded wire reinforcing (WWR) sizes. The lap splice length is sometimes referred to as the development length. When lapping bars in sections where the moment is greatest, it is important not to lap more than 50% of the reinforcing at these critical sections.

8.2 Design Capacity and Reinforcing Requirements

The shear and moment capacities of a beam section based on strength or limit design principles or load resistance factor design (LRFD) are presented in this section.

Table 8.1 Reinforcing steel – concrete cover from AASHTO 5.12.3.1.

	Cover (in)
Cast against earth	3
Exterior other than above	2
Interior	1.5
Bottom of cast-in-place slabs	1

The design capacity of a section, referred to as the factored resistance, equals the nominal capacity times a resistance factor, f, presented in Table 8.2 for concrete members. This design factored resistance must be greater than the applied factored loads.

8.2.1 Shear Design Capacity

The design shear resistance capacity is equal to:

$$V_r = \phi V_n$$

For shear the resistance factor is 0.90.
For single shear (AASHTO 5.8.3.3) in terms of stress,

$$v_r = \phi v_n = \phi v_c = 0.90 \times 0.0316 \times 2(f_c')^{0.5}$$

or approximately 100 psi for $f_c' = 3$ ksi.

For two-way action resulting in double shear (AASHTO 5.13.3.6.3),

$$v_r = \phi v_n = \phi v_c = 0.90 \times 0.126(f_c')^{0.5}$$

or approximately 200 psi for $f_c' = 3$ ksi.

The shear force capacity of a section $V = v(bd)$. For large confinement pressures, an equation that incorporates the increase in shear capacity with confining pressure should be used. For v_{single} the critical section is at a distance equal to member thickness from face of concentrated load (AASHTO 5.13.3.6.1).

8.2.2 Moment Design Capacity

As stated previously, if concrete is unconfined, the maximum usable strain at the extreme concrete compression fiber is approximately 0.003. In concrete members the design capacity of the section is the nominal capacity times a resistance factor, ϕ, presented in Table 8.2. The nominal (ultimate) load capacities used for strength design analysis are when the component materials reach their ultimate strength ($f_s = f_y$,

Table 8.2 Resistance factors ϕ from AASHTO (5.5.4.2).

Strength limit state	ϕ	
Flexure and tension	0.9	
Shear and torsion	0.9	
Shear in buried structures	0.85	(AASHTO 12.5.5-1)
Axial compression	0.75	

$\varepsilon_s > \varepsilon_y$, maximum extreme concrete compression fiber strain, $\varepsilon_c = 0.003$). In other words, although $f_s = f_y$, the strain in the reinforcing can be very much greater than ε_y, and that is why there is the following restriction on the spacing (distribution) of the reinforcement (AASHTO 5.7.3.4) to control crack width. Also, the minimum reinforcement in a section is limited to the lesser of (AASHTO 5.7.3.3.2)

$$M_r > 1.2\,M_{cr} > S_c f_r \quad \text{or} \quad M_r > 1.33(M_r)_{req'd}$$

Usually these limitations on moment are applicable for footings or members not highly stressed in moment:

The nominal (ultimate) or resisting load capacities used for strength design analysis are when the component materials reach their ultimate strength ($f_s = f_y$, $\varepsilon_s > \varepsilon_y$, maximum extreme concrete compression fiber strain, $\varepsilon_c = 0.003$ (AASHTO 5.7.2)):

$$M_r = \phi M_n = 0.9\,M_n$$

For moment, the critical section is at the face of the wall (AASHTO 5.13.3.4).

The ultimate or strength limit design bending moment capacity of reinforced concrete sections are computed in workbook **Beam-LFD**.

8.2.3 Beam–Column Capacity

The capacity of a beam–column is developed using a P–M (compressive axial load versus bending moment) interaction diagram. A P–M diagram shows that there is an increased moment capacity due to precompression of the section. Any P–M combination within the outlined polygon is allowable, but outside it is unacceptable. Usually, unless the structural element is a column, the stress state on the beam–column sections is always in the tension-controlled region of the P–M interaction diagram. The workbook **Beam-Col ID** develops the interaction diagram for a rectangular beam–column section 1 ft wide.

Program **PMEIX-VBA** (Reese and Allen, 1977; Reese, 1984; Ensoft, Inc., 2011) can also be used to develop such curves. By varying the axial loads input, this VBA program can develop the points on the curve of a P–M interaction diagram. The program computes the moment capacity, M, of a section under an imposed rotation, ϕ, and then determines the EI of the section from the equation $EI = M/\phi$. It incorporates the nonlinearity in the concrete's stress–strain relation, E, and the cracking of concrete below the neutral axis causing a reduction in the moment of inertia, I.

The VBA macro program is attached as a set of VBA macro procedures to the workbook **PMEIX-VBA**. The VBA macro is started in Sub PMEXI(). **Example 8.1 (Rect8thk)** shows an application of its use for a rectangular concrete section 8 in thick with #5 bars at 12 in on center subject to a 6.2 k/ft axial load. A plot of PMEIX data for the decay of EI with increasing values of ϕ is given in Figure 8.1a, for the increase in M_d and f_s with increasing ϕ values in Figure 8.1b, and for the decay of EI with increasing M_d values in Figure 8.1c. A comparison of results using programs **PMEIX-VBA,Beam-Col ID**, and **Beam-LFD** on the same section of **Example 8.1** is presented in Table 8.3.

The widely used PCAcol program (PCA, 1992) is a solution for a beam–column. This program has the limitation for circular columns that symmetry of reinforcing about the two axes must be maintained. Thus a 6-bar configuration cannot be symmetrical while a 4-, 8-, 12-, and 16-bar configuration can. Program **PMEIX-VBA** has a couple of advantages over PCAcol. One is that, by not being just a true "column" program, the quantity of reinforcing steel can be different on the tension and compression side for rectangular sections. As such it can be used to compute the capacity of a doubly reinforced section whether in compression or not. Another advantage is that, as the operating system changes from Windows XP (32 bit) to Windows Vista or 7 (64 bit), the programs are not out of date but compatible with either operating system since Excel has been developed to function in both systems.

8.2.4 Shrinkage and Temperature Reinforcement (AASHTO 5.10.8)

Reinforcement for shrinkage and temperature stresses shall be provided near **surfaces** of concrete exposed to daily temperature changes and in structural mass concrete. The area of reinforcement in each direction shall not be less than

$$A_s \geq 0.11A_g/f_y = 0.0018A_g$$

for 60 ksi yield steel. The steel shall be equally distributed on both faces. However, for members 6 in or less in thickness, the steel may be placed in a single layer. This reinforcement shall not be spaced further apart than three times the component thickness or 18 in.

8.2.5 Reinforcement to Control Cracking

To control live load deflection and limit cracking for serviceability the following three provisions are suggested by the AASHTO code:

1. Limit for serviceability the LL deflection of the structure under a design truck (AASHTO 3.6.1.3.2) with 32 k axle loads to Span/800 (AASHTO 2.5.2.6.2).
2. Use the equations of Article 5.7.3.4 AASHTO Control of Cracking by Distribution Reinforcement. By the equations in this article the crack widths are limited to 0.017 in for Class 1 (interior face) exposure conditions and to 0.1275 in for Class 2 (exterior) exposure conditions and to 0.0085 in for severe or buried exposure. This calculation insures that the steel is not placed so far away from the tension face that cracking would become excessive in the concrete before the section has failed.

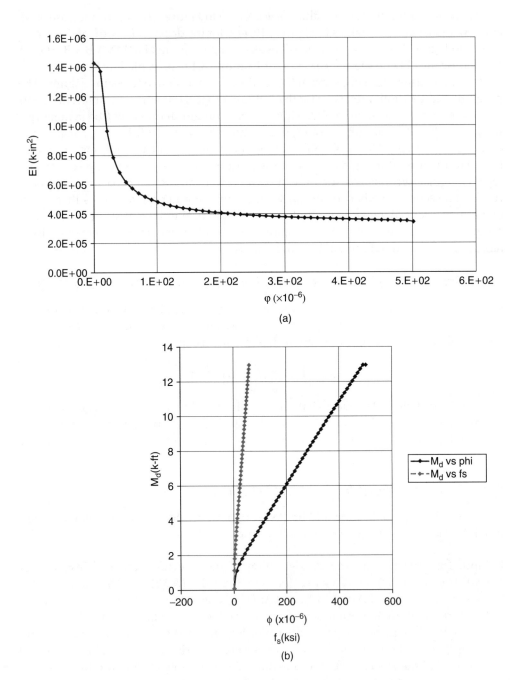

Figure 8.1 (a)–(c) PMEIX data plot of Example 8.1 – 8″ thick section.

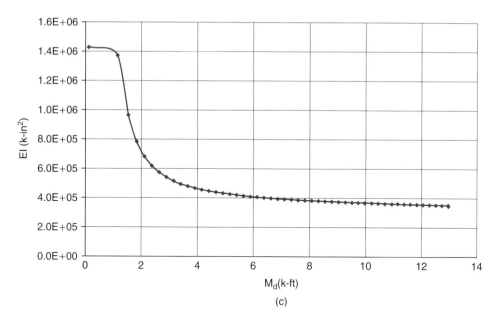

Figure 8.1   *(continued)*

Table 8.3   Comparison of program results.

Program	I_{cr}/I_g	M_d(k-ft)
PCA formula	0.20	–
PMEIX[a]	0.22	13
Beam-LFD	–	11.9
Beam-Col ID	–	12.8

Data for Example 8.1.
[a] Assuming no change in E.

3. Provide a quantity of reinforcing as per Article 5.6.3.6 AASHTO Crack Control Reinforcement such that the reinforcement supplied is not less than $0.003\,A_g$ in each direction.

8.3 Strength Properties for a Soil–Structure Interaction Analyses

In a frame analysis, it is common to model the structure as being linear elastic with the area and moment of inertia of an uncracked concrete section I_g. The relative stiffness values of the various component sections of the frame are important in distributing the load through the structure. When the structure is supported by or working with the surrounding soil to carry and distribute the applied load, the ratio of the stiffness of the structure to that of the soil is important in allocating the load carried by each. However, for such an analysis it may be desirable to perform the analysis using cracked section properties for the structure. In such an analysis, though, it is often difficult to assess which portions of the structure, if any, will crack.

Article 4.5.2.2 AASHTO Elastic Behavior states that: "The stiffness properties of concrete and composite members shall be based upon cracked and/or uncracked sections consistent with the anticipated behavior." Design of structures at the strength limit state requires that the exact state of stress in each member be ascertained in order to determine what part of the structure is failing. This is similar to a sequential plastic analysis where the structure is loaded incrementally to ascertain which individual members are progressively loaded into the plastic state.

A lower limiting condition evaluation of the structural system would result if $(EI)_{cr}$ is used for all members and then the structure is checked to see that it is stable (remains standing). As long as the strength curve is still strain hardening (not softening), the structural configuration remains stable.

8.4 Cracked-Section Concrete Properties

The strength as governed by a section's moment capacity is $M_{cap} = f'_c I/y$. The strength capacity is related to the moment of inertia, I, and not the elastic modulus, E, of the section. The bending stiffness of a concrete section is defined by the relation $\phi = M/(EI)$. EI decreases as the section is deformed by increasing its curvature, ϕ, up to the section's moment capacity in the strength limit state. For concrete, E is relatively constant throughout its load range. Thus the parameter I generally governs the section's strength and stiffness.

To insure safety against structural failure, all sections shall be designed for the applicable factored loads specified at the strength limit state. At this limit state, the steel reinforcing exceeds its initial yield strain and considerable cracking occurs, causing a reduction in the moment of inertia even before the maximum usable strain of 0.003 is attained at the extreme concrete compression fiber. Cracking of concrete below the neutral axis occurs and causes a reduction in the section's moment of inertia, I. At the strength limit state, where a cracked section exists, a modification to the parameter I may be applicable.

I_{cr} determinations can be developed from any of the following three sources:

1. Equations presented in Table 8.2 on page 8-4 of PCA Notes (1996).
    Using the defined parameters: $n = E_s/E_c$ $B = b/(nA_s)$
    For single reinforcement:

$$kd = \{(2dB + 1)^{0.5} - 1\}/B \quad I_{cr} = b(kd)^3/3 + nA_s(d - kd)^2$$

For doubly reinforced section: $r = (n-1)A_s'/(nA_s)$

$$kd = [\{2\,dB(1 + rd'/d) + (1+r)^2\}^{0.5} - (1+r)]/B$$

$$I_{cr} = b(kd)^3/3 + nA_s(d-kd)^2 + (n-1)A_s'(kd-d')^2$$

These equations are employed in the workbook **Beam-LFD**.

2. Equations (10-10), $(EI)_{cr} = E_cI_g/5$, and (10-11), $E_cI_g/2.5$, presented in Article 10.11.5.2 of the ACI 318-89 Building Code (1989). E_c is the concrete modulus of elasticity, and I_g refers to the gross concrete section. These equations specify a reduced EI to represent slenderness of an element in a buckling analysis. Similar equations for approximate evaluation of slenderness effects are presented in Article 5.7.4.3 of the AASHTO LRFD 2010 Bridge Design Specification.

   At lower reinforcement ratios, approximations for the moment of inertia for a cracked section of $E_cI_g/2.5$, suggested by Equation (10-11) in the ACI 318-89 Building Code, Section 10.11.5.2, may be more like the $E_cI_g/5$ to $E_cI_g/10$ of Equation (10-10) of that reference.

3. Program **PMEIX** developed by Reese (1984).

   Both the stiffness and moment capacity of circular and rectangular sections can be computed using this program. This program computes the moment capacity, M, of a section under an imposed rotation, ϕ, and then determines the EI of the section from the equation $EI = M/\phi$. It incorporates the cracking of concrete below the neutral axis that causes a reduction in the moment of inertia, I, into the moment capacity of the section.

   Program **PMEIX** shows that an 8 in thick concrete section doubly reinforced with #4 @ 8 in and loaded with an axial compressive force of 10 k/ft deforms to the point where the steel first yields (at which point the maximum concrete strain is only 0.0008). For this section $M_{cap} = 9.9$ k-ft, and EI changes from 16×10^5 in^4/ft to 2.8×10^5 in^4/ft or to 18% of the uncracked value (approximately a factor of 6). See Figure 8.1 showing this behavior. This factor of 6 is exactly predicted by cracked moment of inertia calculations specified by PCA.

   Such a change occurs predominantly in the section modulus (geometry–moment of inertia) and not the elastic modulus (material). At this level of concrete strain there is still more capacity and strength to be derived in the section before it reaches the strain level of 0.003. Even at the level of the working stress moment, $M_{cap-ws} = 5.3$ k-ft, the extent of cracking in the section is similar to what exists at the limit state, with the cracked moment of inertia being 23% of the uncracked value.

8.5 Excel Workbooks

8.5.1 Workbook Reinf Concrete

The stress–strain curve for concrete is shown in worksheet Stress-Strain Prop.

Reinforcing quantities as a percentage of p_b are given as A_s/A_g.

The workbook contains various design parameter relations such as steel reinforcing sizes, A_s/ft, lb/cy, lap splice length, and WWR sizes.

8.5.2 Workbook Beam-LFD

The ultimate or strength limit design bending moment capacity of reinforced concrete sections are computed in this workbook along with cracked-section properties specified in the equations in Table 8.2 on page 8-4 of PCA Notes (1996).

8.5.3 Workbook Beam-Col ID

This workbook develops a point on the interaction $(P-M)$ diagram for a rectangular beam–column section 1 ft wide.

8.5.4 Workbook PMEIX-VBA

This workbook develops a point on the interaction $(P-M)$ diagram for a circular or rectangular beam–column section. **Example 8.1 (Rect8thk)** supplies an input data worksheet for the specific section of a rectangular section beam–column 8 in thick by 12 in wide reinforced with #5 bars at 8 in on center.

 Related Workbooks on DVD

Beam-Col ID(4)
Beam-LFD(2)
Example 8.1 Rect8thk
PMEIX-VBA(5)
Reinf Concrete(2)

8.6 Notation

a = depth of equivalent rectangular stress block = $\beta_1 c$
A_g = gross area of concrete section
A_s = area of tension reinforcement
A_s' = area of compression reinforcement
b = width of compression face of member
c = distance from extreme compression fiber to neutral axis
d = distance from extreme compression fiber to centroid of tension reinforcement
d' = distance from extreme compression fiber to centroid of compression reinforcement
d_b = diameter of reinforcing bar

E_c = modulus of elasticity of concrete
E_s = modulus of elasticity of steel reinforcement
EI = flexural stiffness of compression member
f_c' = compressive strength of concrete
f_y = yield strength of reinforcement
I_g = moment of inertia of gross concrete section about centroidal axis, neglecting reinforcement
l_d = development length
M_u = factored moment applied to section of member
M_n = nominal moment resistance
M_r = factored moment resistance = ϕM_n
$n = E_s/E_c$
$p = A_s/A_g$
v_c = nominal shear stress resistance of concrete
v_r = factored shear stress resistance = ϕV_n
v_n = nominal shear stress resistance
V_c = shear stress resistance of concrete
V_r = factored shear resistance = ϕV_n
V_n = nominal shear resistance
$w = \gamma$ = unit weight
α = coefficient of thermal expansion
β_1 = stress block factor = 0.85 for $f_c' \leq 4$ ksi
ϕ = strength reduction factor in equation design strength = ϕ nominal strength
$\phi = M/(EI)$ = rotation or curvature of section in bending
$\rho = A_s/(bd)$
$\rho' = A_s'/(bd)$
$\rho_b = A_s/(bd)$ = reinforcement ratio at balanced strain conditions when the maximum strain at the extreme compression fiber reaches 0.003 simultaneously with the yield strain in the tension reinforcement.
ν = Poisson's ratio

References

AASHTO LRFD (1994) *Bridge Design Specifications*, 1st edn, (1998) 2nd edn, (2004) 3rd edn, (2007) 4th edn, (2010) 5th edn, American Association of State Highway and Transportation Officials, Section 5 Concrete Structures.

ACI (1989) *Building Code Requirements for Reinforced Concrete (ACI 318-89) and Commentary – ACI 318R-89*, American Concrete Institute.

Ensoft, Inc. (2011) LPILE Plus 5.0 Contains PMEIX.

Ghosh, S.K., Fanella, D.A., and Rabbat, BG (eds) (1996) *Notes on ACI 318-95, Building Code Requirements for Structural Concrete with Design Applications*, Portland Cement Association.

PCA (1992) *Program PCAcol*, Portland Cement Association, Program PMEI Instructions and Listing, pp. 189–224.

Reese, L.C. (1984) *Handbook on Design of Piles and Drilled Shafts Under Lateral Load*, FHWA-IP-84-11, University Press of the Pacific, July.

Reese, L.C. and Allen, J.D. (1977) *Drilled Shaft Manual*, Structural Analysis and Design for Lateral Loading, Vol. II, USDOT, FHWA, IP 77-21, July.

USACE (1992) *Strength Design for Reinforced-Concrete Hydraulic Structures*, Engineer Manual EM 1110-2-2104, US Army Corps of Engineers, June 30.

Part Three
Soils

"One reason why mathematics enjoys special esteem, about all other sciences, is that its laws are absolutely certain and indisputable, while those of other sciences are to some extent debatable and in constant danger of being overthrown by newly discovered facts."

– Albert Einstein

"Imagination is more important than knowledge."

– Albert Einstein

9

Soil Classification

9.1 Field Geotechnical Processes

Soil and rock strata are explored by sampling them and testing them while they are beneath the surface. The field engineer applies descriptions to the logs he or she creates from these results.

9.1.1 Soil/Rock Exploration

Auger boring investigation and sampling (ASTM Standard D1452) are often conducted using drill rigs capable of auger drilling, rotary wash drilling, and rock coring. For difficult drilling conditions, tricone or Tubex methods must be employed. Where the rock is competent it can be cored. Applicable ASTM standards for field and laboratory testing are presented in Table 9.1.

9.1.1.1 Auger Boring

Auger drilling is typically performed using $7\frac{1}{4}$ in o.d. $\times$ $3\frac{1}{2}$ in i.d., or $6\frac{5}{8}$ in o.d. $\times$ $3\frac{1}{4}$ in i.d. hollow-stem or 4 in o.d. solid-stem augers, both with carbide-tipped teeth. Samplers are attached to a rod and extend down to the bottom of the hole through the hollow portion of the auger. Alternatively, sonic drill rigs drive a core barrel using vibratory methods and extract the encapsulated soil sample.

9.1.1.2 Tricone Gear Bit

An alternative method for strata that cannot be penetrated using hollow-stem auger drilling is a tricone approach. Rotary wash drilling employs a tricone gear bit attached to a rod extending through the center of the auger. The cuttings may be removed using either high-pressure air or water, with or without admixtures, as a drilling fluid. This fluid also lubricates the bit head.

Solutions for Soil and Structural Systems using Excel and VBA Programs, First Edition. Robert L. Sogge.
© 2012 John Wiley & Sons, Ltd. Published 2012 by John Wiley & Sons, Ltd.

Table 9.1 Applicable ASTM standards.

Drilling test	ASTM standard
Auger boring investigation and sampling	D1452
SPT and split-barrel sampling	D1586
Thin-walled tube sampling	D1587
Laboratory test	**ASTM standard**
Classification and description	D2487 and D 2488
Gradation	D6913 and D422
Sieve – minus #200	D1140
Atterberg limits	D4318
Natural moisture content and dry density	D2216 and D2937
Compaction – modified proctor	D1557
Compaction – standard proctor	D698
Direct shear (three points)	D3080
Unconfined compressive strength	D7012
Point load test	D5731

9.1.1.3 Down-Hole Hammer Drilling with Casing Advancement

This type of drilling, known as the Tubex/Concentrix drilling system, is ideally suited for situations where there is a need to drill and case a hole in formations where the advancement of a hollow-stem auger is not possible and the formation is not suitable for coring. This system incorporates a down-hole hammer (DHH), a pilot bit, a reamer, and a hardened steel shoe on the lead casing. The drill string and hammer are placed within the casing, revealing the pilot bit and eccentric reamer. High-pressure/volume air pressure (300 psi at 750 cfm) is used to actuate the hammer and facilitate the removal of drill cuttings. When the drill string is rotated to the right, the eccentric reamer is deployed and cuts a hole of a slightly larger diameter than that of the casing. As the hammer drills, it pulls the casing down simultaneously. Drill cuttings are lifted between the casing and the drill string and discharged away from the hole by way of a diverter. Once the target hole/sample depth is achieved, drive samples can be collected or wells installed in a method consistent with hollow-stem auger drilling.

9.1.1.4 Diamond Core Bit

For drilling in rock a diamond drill bit is attached to a core barrel. Barrels can be extended down the center of a hollow-stem auger and essentially use the auger as an outer casing that contains the drilling fluids. The hole is advanced by the downward pressure on the drill rod that is attached to the core barrel. The diameters of rock cores generally range from a little less than 1 in to over 3 in.

9.1.2 Soil/Rock Sampling

Sampling of the soil and rock is required for complete identification and testing.

9.1.2.1 Soil Sampling

"Disturbed" samples are taken using a standard 2 in o.d. $\times$ $1\frac{3}{8}$ in i.d. split-spoon sampler (ASTM Standard D1586). The i.d. dimension is for the o.d. of any inner brass liner. "Undisturbed" samples of cohesive fine-grained soils are obtained using a ring sampler of 3 in o.d. $\times$ 2.5 in i.d. often referred to as a Dames & Moore sampler. In very soft clays, thin-walled Shelby tubes (ASTM Standard D1587) can be pushed into the soil. A series of 1 in long $\times$ 2.416 in i.d. brass rings in the ring sampler have a 2.5 in o.d. and therefore readily fit into laboratory direct-shear and consolidation equipment. Alternatively a Dennison barrel sampler can be used to obtain samples in formations that are too soft to core and too hard to push in Shelby tubes or drive in ring samples.

Another way to obtain undisturbed soil samples is with a continuous sampler barrel that rides inside the hollow-stem auger. It looks like an oversized sampler that is 5 ft long and just smaller than the inside diameter of the augers. It projects a shoe just ahead of the auger bit and gives a continuous core of the soils. The sampler is connected to a bearing assembly up above the drive head that keeps the barrel from turning with the augers.

9.1.2.2 Rock Sampling

The core barrel is used to contain the rock core sample. Such barrels can be single-, double-, and triple-tube core barrels. In each of them the outer barrel has a cutting head section with a diamond bit attached. The sample can be extracted using a wireline that extends to the core barrel for retrieving the core sample.

9.1.3 Field Testing

Testing or describing the characteristics of the in-situ soil or rock material is done by the field engineer.

9.1.3.1 In-situ Soil Testing

An approximation of the soil's in-situ density and consistency, from which an estimate of the soil's strength estimates can be made, is obtained using the penetration resistance to driving of the samplers. The standard penetration test (SPT) N value is the number of blows to drive a standard 2 in o.d. $\times$ $1\frac{3}{8}$ in i.d. split-spoon sampler 1 ft using a 140 lb weight dropping 30 in. Where driving resistance is high, blows per inch driven are recorded. A 3 in o.d. ring sampler will generally have a higher blow count than will a split-spoon sampler for the same penetration distance because its frontal penetration area is greater. Continuous penetration resistance can be measured with a 2.0 in o.d. bull-nose penetrometer. When using the SPT driving energy the blow counts on a bull-nose penetrometer are approximately equal to or greater than the SPT N value. As an alternative to SPT, cone penetration testing (CPT) can be performed using specialized driving equipment to assess the density of the in-situ soil.

9.1.3.2 Rock Characterization

Where a tricone or Tubex method is used, a characterization of the competency of the material is obtained from the rate of hole advancement. Where coring of the material is applicable, the rock is sampled in 5 or 10 ft-long core runs that are stored in a core box. The rock is logged with the boring number, depth of sample, and rock type. The amount and the condition of the recovered rock core are denoted as the core recovery (CR) and the rock quality designation (RQD) respectively. CR is the percentage of the length of the core run that is recovered. RQD expresses the amount of fracturing and joints in the rock and is defined by the equation

$$RQD = \Sigma \text{ pieces} > 100\,\text{mm (4 in)/length of core run}$$

RQD is expressed as a decimal.

9.2 Soil Description

Complete soil descriptions consist of the following parts:

1. Color of the soil.
2. Basic soil type.
3. Modifying terms.
4. Identification of special soil types.

9.2.1 Color

The simplest and least ambiguous descriptive terms are those of the primary and binary colors modified by the adjectives light or dark. Medium shades require no modifying adjective. Terms like mauve, beige, orchid, tan, should be avoided.

9.2.2 Basic Soil Type

The terms boulder, gravel, sand, silt, and clay are used to demarcate soil gradations. Commonly, the following grain-size definition of soil components is used:

Boulders – larger than 6 in
Cobbles – 2–6 in
Gravel – 2 mm to 2 in
Sand – 0.074–2 mm
Silt – 0.002 mm (2 μm) to 0.074 mm
Clay – smaller than 0.002 mm (2 μm).

The given demarcations refer only to the particle dimensions and do not imply anything about the mineral composition, geologic origin, or soil forming process. For example, particles of clay size may be true clay minerals, other layer silicates (such as mica), or rock flour having almost any mineral composition. In a strict sense, individual particles or groups of particles in the fine size range require further property definitions to define them specifically as a silt or clay.

In general, use of the term rock should be avoided in soil description because it is ambiguous. It should be further defined as the type of rock and, if particles are larger than gravel size, as rock fragments or bedrock. Some older classification systems that were derived from the US Department of Agriculture, Bureau of Soils system, make use of the term loam. Modern systems do not use this term because it carries the connotation that the soil contains humus or other organic matter. Because the upper organic soil horizons are never used for engineering purposes, a better term for it on boring logs is topsoil.

9.2.3 Modifying Terms

The noun in a soil description is always the most abundant component and the adjective is always the component next in quantity. Thus, there are clayey sands, sandy clays, silty sands, clayey silts, and so on. The adjective is for describing the name and quantity of the secondary component and is stated as:

Trace for <12%
Some for 12–30%
"ly" for >30%.

Modifying terms can be modifying clauses which follow the basic soil type, and adjectival modifiers which precede the color and basic soil type.

Examples of the first type are:

silty clay with some fine gravel;
sandy gravel with trace cobbles;
angular coarse brown sand; and
varved blue silty clay.

Examples of complete soil descriptions follow:

rounded light-brown medium sand with some fine gravel;
black organic muck with some clay;
varved blue silty clay with some fine sand partings, sub-angular;
gray clayey gravel with a few scattered cobbles and an occasional boulder.

9.2.4 Special Soil Types

Certain types of soils have sufficiently distinctive characteristics that they may be identified immediately by their unique properties. Some of these soil types predominate in certain parts of the United States: for instance, caliche is widespread throughout the arid west and southwest; marl is common in areas bordering the central and eastern parts of the Gulf of Mexico.

Other soil types like peats and organic mucks, while widespread in occurrence, are found in relatively small local deposits. Peat can be distinguished by its color (brown to black), by the presence of a considerable amount of partly carbonized plant fiber, by its odor (a swampy smell), and by its very high compressibility. When dry, it is brittle and can usually be ignited. Dry peat will float in water. Peat may contain some very fine sand, silt, and clay as well as organic matter. Organic muck is similar to peat except that it has little or no fibrous material, usually has more inorganic sand, silt, and clay content, and usually cannot be ignited when dry.

Caliche is a fairly well-graded mixture of sand, silt, and clay (occasionally some gravel) which has been lightly to completely cemented by water-deposited calcite (lime). When dry, it may vary from crumbly and chalky to very hard and rocklike. It has a characteristic white to very light-brown color and will effervesce strongly when a drop or two of dilute hydrochloric acid is placed on it.

Marl is a very fine clayey sand to sandy clay having a high calcium carbonate content which is disseminated throughout the soil and which usually has not produced any noticeable amount of cementation. It can be identified by its appearance and by strong effervescence when treated with dilute hydrochloric acid.

Bentonite is a fine-grained, highly swelling colloidal clay. It can be identified by the fact that air-dry pieces are very hard, and, when submerged in water, will swell to several times the dry volume, or more, and will have a gelatinous appearance and a soapy feel. When at natural moisture contents greater than air dry and less than saturation, bentonite will have a waxy to soapy appearance and feel. In this state, fracture surfaces and slickensides can usually be seen. When wet enough to be well up in the plastic range, bentonite is extremely sticky. Other, even more swelling clays are illite and montmorillonite.

Topsoil can be distinguished by the presence of roots, rootlets, root holes, worm holes, and a somewhat blocky structure, by the color (usually yellow, red, brown, or black), and by the organic content. The color usually becomes progressively lighter with depth in the stratum.

Fill is the name given to all human-made deposits of natural soil or waste materials. It may consist of almost any conceivable material or mixtures thereof. When the fill is composed of soil materials, identification may be difficult but sometimes can be accomplished by noting the lack of regular bedding and by the presence of topsoil underlying the fill. Fills of non-soil materials or mixtures of soil and non-soil materials usually contain human-made objects, or artifacts, such as brick or concrete fragments, plaster, pieces of wood, metal objects, rubber, and the like.

Loess is an aeolian (deposited by wind) material that, due to its method of deposition, is usually almost uniform in grain size (very fine sand and coarse silt). Particles are angular to sub-angular forming a strong interlock when the soil is dry, which strength is lost completely when disturbed and remolded. Loess will stand in a stable configuration in vertical cuts, but the material will collapse on wetting at less steep slope angles.

9.3 Field and Laboratory Tests for Soil Identification

In order to develop the soil description to be placed on the field logging sheet, soils may be tested in the field. The description of disturbed soil samples collected from split-spoon samples as well as field cuttings and undisturbed soil samples collected from ring samplers consists of soil color, basic soil type as a boulder, gravel, sand, silt, or clay, and modifying terms.

Accurate field identification of soil types and accurate, complete descriptions of soils encountered in the field are a necessary and very helpful part of any soil investigation. This information forms the basis for preliminary screening of field samples before using laboratory tests to confirm the identification. Soil descriptions appear on field logs and are based on the field identification by a field soils engineer. These identifications are accomplished by visual examination supplemented by a few simple field tests that have been performed quickly and without elaborate equipment. Good field descriptions reduce the amount of laboratory testing for this purpose that may be needed to prepare final boring or test-pit logs, and aids in extrapolating boring profiles to soil profiles over the investigated area. Incomplete or erroneous field identification always multiplies the amount of laboratory work required and may lead to misinterpretation of the soil profile. Laboratory testing of selected soil samples must be conducted in order to make sure that the initial descriptions given to the soil in the field are accurate.

9.3.1 Field Tests for Soil Identification

Simple identification tests that can be conducted by the engineer in the field can identify soil types. Such tests are mainly applicable to fine-grained soils. In many cases, no single test or examination will be completely diagnostic, but several will usually identify the basic soil type. There are always some "borderline" soils which will be difficult to identify, such as silty sands or sandy silts or silty clays or clayey silts. In these cases the description which is the more coarse grained will usually be correct. Coarse-grained soils are rarely identified incorrectly because all of the grains can be discerned with the unaided eye. The lower limit in grain size of 0.074 mm for fine sand is about the limit of visual resolution for the average eye. Below this size individual grains cannot be distinguished; hence other means than visual examination must be used.

The application of simple field tests for the examination and identification of soils is as follows:

1. **Coarse-grained soils**. Identification of all coarse-grained soils in either the wet or dry state is accomplished by visual examination of grain size, grain shape, and gradation.
2. **Fine-grained soils**. In the identification of fine-grained soils the simple field identification tests 2–6 below are used to distinguish between silt and clay; or, in the case of mixed soils, to roughly determine the relative percentages of sand, silt, and clay-size particles.

These field identification tests are as follows:

1. Visual examination of grain size and grain shape for coarse-grained soils only.
2. Molding test – estimation of plasticity by molding with the fingers.
3. Crushing test and dusting test – estimation of cohesion in the dry state.
4. Molding test – to yield change in consistency and strength during remolding with the fingers.
5. Shaking test – the pore water mobility and dilatancy can be observed by shaking in the hand and then shearing.
6. Settling test – from observation of the rate of sedimentation and character of sediment and suspension of a small amount of soil in water shaken in a test tube and allowed to settle.

As a starting point in the identification of fine-grained soils, it is usually helpful to determine first whether the soil is organic or inorganic. Organic content is usually indicated by a dark-gray, brown, or black color, sometimes showing banding, and often having a faint to strong odor of decay. When molded (molding test), organic silts and clays have a noticeably softer feel to them than corresponding inorganic soils. Organic content increases the plasticity and stickiness (molding test) but decreases the dry strength (crushing test).

The next step is to determine whether the soil is predominantly silty or clayey. In this determination, the most reliable indicators are the crushing and dusting tests, although others may be useful in confirming the identification. In the crushing test, a small piece of the soil (undisturbed state preferred) is air dried and then crushed between the fingers. Silts will crumble and powder under light to moderate finger pressure, whereas clays will fracture or break without crumbling or powdering and will require strong finger pressure. Some clays may be strong enough to resist fracture entirely.

The dusting test is performed by first making a thin smear of very wet soil on the topside surface of the hand, then allowing it to dry, and finally attempting to brush the soil off using the other hand. If the soil is mostly silt, it will dust off readily leaving the hand reasonably clean. If mostly clay, the soil will resist brushing off and scales, rather than dusts, as it comes off the hand. Usually, clays strongly resist removal in this manner and quite commonly the hand will be stained.

The shaking test will immediately identify very fine sands, coarse silts, and rock flours. In this test, a small amount of the soil is thoroughly mixed with water to form a fairly stiff saturated slurry. The mixture is shaken rapidly from side to side and the observation made as to whether or not pore water is brought to the surface (pore water mobility). Very fine sands, coarse silts, and rock flours will become quite shiny; clays will show little or no change in appearance. The wet soil is then sheared and the surface appearance noted during shear. If it becomes dull, indicating an uptake of surface water, the soil is very fine sand, coarse silt, or rock flour. Clay will not change its surface appearance during shear. Uptake of water during shear is also accompanied by a noticeable increase in resistance to shear (dilatancy).

The molding test can be used to aid in distinguishing between silt and clay by rolling the soil out into threads and observing whether or not the threads have tensile strength. If silt, the thread will break into pieces when it is picked up by one end. If clay, the thread will remain intact when picked up and, in addition, will show some tensile strength when pulled apart. The relative amount of tensile strength will indicate whether the clay is lean or fat.

The molding test can also be used to determine whether or not a clay is sensitive or has a difference in strength between the un-remolded and remolded state. A small piece of the wet undisturbed soil is crushed between the fingers and the relative amount of effort required is noted. The soil is then thoroughly remolded and formed into a rough cube and again crushed between the fingers, again noting the relative effort required. If little difference in crushing strength is noted, the clay is insensitive; if there is a noticeable difference, the clay is sensitive. The degree of sensitivity depends upon the magnitude of the difference.

The settling test will determine the approximate amounts of each component for soils where sand, silt, and clay-size particles seem to be present in approximately equal amounts. A small amount of dry powdered soil is shaken with distilled water, to which a dispersant has been added to prevent flocculation, for about a minute in a 4 in test tube and then set aside in a rack to sediment out. All sand-size particles will settle out in less than 30 seconds. Silt-size particles will require 20–25 minutes to completely settle out. At the end of 2 hours, all but the finest colloidal clay will have settled out. At that time, the sediments in the bottom of the test tube will be observed to have reasonably well-defined layers of sand, silt, and clay that can be estimated from the observed thickness of the layers. Because the silt will be unconsolidated and almost liquid, the height of the silt layer should be divided by 2 for this comparison.

9.3.2 Laboratory Testing for Soil Identification

Laboratory testing is required to confirm the soil description given to the soil initially in the field. Such tests are for grain size, gradation, and Atterberg limit determination. For testing and classification of soils see AASHTO, ASTM, ASTM Committee D-18, Bowles (1970), Germaine and Germaine (2009), Lambe (1951), Lambe and Whitman (1969), Portland Cement Association (1972) and Taylor (1948).

9.3.2.1 Gradation

After a soil sample received from the field has been prepared by selecting representative portions, a mechanical sieve analysis for a true gradation determination and consistency limit tests can proceed. The gradation test (ASTM Standard D6913) is performed on material coarser than a #200 sieve that has 0.074 mm openings. The fine-component gradation is developed using a hydrometer test (ASTM Standard D422). Usually this latter time-consuming test is not required and a #200 sieve wash (ASTM Standard D1140) will suffice.

9.3.2.2 Atterberg Limits and Indices

Tests to determine consistency or Atterberg limits (ASTM Standard D4318), sometimes referred to as consistency limits, are performed on material passing a #40 sieve. Terms associated with these limits and various other indices are presented in the Appendix to this chapter.

For design purposes additional tests on the soil must be conducted to provide, for example, the suitability of materials for subgrade or sub-base construction or to yield design information on thickness of pavement and sub-base required to support anticipated wheel loads or to yield information on a soil's structural and performance characteristics.

Such tests that can be prescribed are for compaction, permeability, strength, and settlement consolidation or swell.

9.4 Soil Classification Systems

Classification systems for expressing the results of these tests are based on two approaches:

1. Textural classifications that rely solely on the distribution of grain sizes obtained from a sieve analysis.
2. Engineering classifications that are based on both gradation from a sieve analysis and Atterberg or consistency limits.

9.4.1 Textural Classification

The preliminary classification of soils in most laboratories is made by utilizing the results of the mechanical sieve analysis to classify the soils according to the gradation of the soil grain sizes. Such a classification is termed a textural classification. The division into size fractions is consistent with those given previously for field identification of soils.

9.4.2 Engineering Classification

Classification by means of grain-size distribution alone (textural) will give little information on the structural or performance characteristics of the soil but will frequently indicate what additional tests must be run in the laboratory to define these characteristics. Determination of soil behavioral characteristics should include the consistency limits of the soil fines. The Unified Soil Classification System (USCS) (ASTM Standard D2487) and its description (ASTM Standard D2488) is the most commonly used soil classification system that includes the consistency limits in its classification. The USCS is based upon soil characteristics which are indicative of the soil behavior as a foundation or construction material. Table 9.2 outlines the procedures for classification of 15 soil types. It should be noted that the system can be used for either field identification

Table 9.2   Unified Soil Classification System.

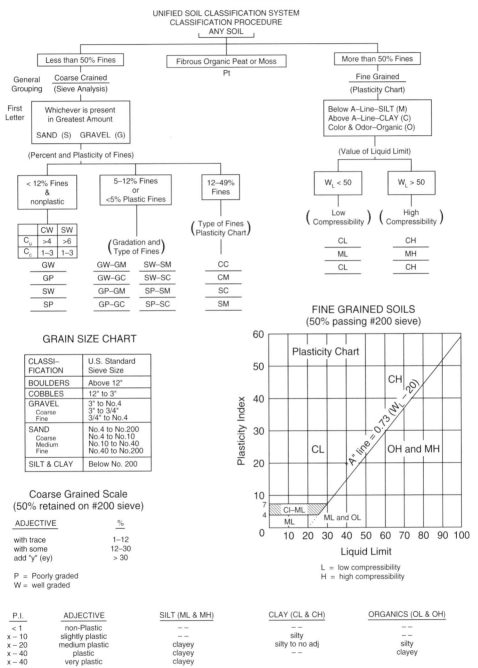

or laboratory classification. Field identification procedures, in general, parallel those detailed for the field identification of soils given previously in Section 9.3.1. Laboratory criteria for classification depend upon data secured from the mechanical sieve analysis of soils and the consistency limits of soils.

In the USCS, soils are divided into two broad categories of coarse-grained soils and fine-grained soils. Coarse-grained soils grade from essentially clean gravels (GW) to sands with appreciable amounts of fine-grained binder (SM, SC). Fine-grained soils grade from coarse silts with slight plasticity (ML) to clays with high plasticity or high organic content with medium to high plasticity (CH, OH). Highly organic soils which do not fit into the two major categories, such as peat and organic muck, are placed in a separate category (Pt) and are identified by color, odor, low density, high water content, extreme compressibility, and presence of carbonized plant fiber rather than by gradation and plasticity.

In the USCS, soils are classified in the laboratory on the basis of:

1. Relative percentages of gravel, sand, and fines (fraction passing the # 200 sieve).
2. Shape of the grain-size distribution curve (as characterized by the coefficient of uniformity (C_u) and the coefficient of gradation (C).
3. Soil plasticity characteristics (Liquid Limit and Plasticity Index).

The coefficients of uniformity and gradation are defined in the table on the basis of D_{10}, D_{30}, and D_{60} which are the grain sizes corresponding to 10, 30, and 60% passing, respectively, on the grain-size distribution curve. The relationship between plasticity and classification is shown on the Plasticity Chart. Referring to this chart, clays (C) generally plot above the "A" line and silts (M) plot below the "A" line. The silt and clay groups are further subdivided on the basis of low (L) or high (H) Liquid Limit that is associated with relative compressibility.

 ## 9.5 Excel Workbooks and VBA Programs

The workbooks and programs associated with this chapter classify soil using the USCS. The workbooks containing the embedded VBA macro programs are based on a BASIC program presented by Stevens (1982) in *Civil Engineering*. Various commercial programs integrating classification with other soil laboratory computations are available to perform this task.

The VBA macro program, **Soil Class-VBA**, employs VBA code containing IF statements and branching. An alternative Excel workbook, **Soil Class**, provides an example of using worksheet IF statements to replace those in the VBA macro and eliminate the need for the VBA macro.

In the VBA macro program version the input data is read from and output data is written to the Excel worksheet, rather than a sequential data file. In such cases it is necessary to clear out data in those cells which may not receive overwriting data in the next run.

The two workbooks **SoilClassUsingVBA** and **Soil Class-VBA** show examples of storing macros in two different locations. A macro containing VBA code is a Sub Procedure that can be inserted by the VBA Editor into a workbook as either:

(1) an Excel Object in Sheet1 (worksheet name on bottom tab) (**Soil Class-VBA**); or
(2) a Module (**SoilClassUsingVBA**).

Both of these storage locations are visible through Project Explorer.

If the macro is put into the worksheet Excel Object, visible by bringing up Project Explorer in the VBA Editor menu under View, one macro is apparent when clicking on Sheet1 and the other is apparent when clicking on Module. In this case both macros are the same and have the name of the Sub, CLASS.

The name of a macro attached to a workbook can be identified through Tools > Macro > Macros. A Sub Procedure can be identified as originating from All Open Workbooks, This Workbook, or a specific workbook name.

VBA code can be copied and pasted into:

1. This "Workbook" under Microsoft Excel Objects of Project Explorer.
2. This "Sheet1" under Microsoft Excel Objects of Project Explorer.
3. A Module.

Typically in this book the VBA programming code is stored in a Module and the VBA macro is started using the familiar form of command button denoted as a "Start" button in the worksheet. By storing code in a module, naming consistency results. Storage in either a worksheet or workbook is applicable to this button triggering the initiation of the VBA code. This text will deal only with code stored in a module or a worksheet and will not address event programming code.

 Related Workbooks on DVD

Soil Class – contains only worksheet IF statements
Soil Class-VBA – VBA macro program stored in Sheet 1
SoilClassUsingVBA – VBA macro program stored in Module

9.6 Soil Mechanics Symbol Nomenclature

Terms Associated with Soil Classification

For grain-size distribution:

$$C_c = \text{gradation coefficient} = (D_{30})^2/(D_{60}D_{10}).$$

C_u = uniformity coefficient = D_{60}/D_{10}.

A well-graded soil will have soil content in each of the component ranges.

A uniform or gap graded soil is denoted as being poorly graded.

D_{10} = the particle size for which 10% of sample is smaller than

D_{30} = the particle size for which 30% of sample is smaller than

D_{60} = the particle size for which 60% of sample is smaller than.

For consistency or Atterberg limits and indices, tests to determine these consistency or Atterberg limits and indices are performed on material passing a #40 sieve.

FL = flocculation limit = level below which soil ceases to display buoyancy.

LL = liquid limit = shear strength of 25 g/cm^2 developed (1 blow = 1 g/cm^2).

Estimate for LL from N value LL = $w_N(N/25)^{0.121}$.

PL = plastic limit.

SL = shrinkage limit = $\{w_o - (V - V_o)/W_s\} = V_o/W_s - 1/G_s$ = the lowest water content at which any further drying will not be accompanied by any further loss in volume.

When the soil changes color from dark to light the sample is at the shrinkage limit.

SR = shrinkage ratio = $(\Delta V/V_o)/\Delta w = W_s/(V_o/\gamma_w)$.

PI = plastic index = LL − PL.

Small PI implies little range of plasticity and the less swelling and creeping that will occur since swelling usually occurs in the plastic state.

LI = liquidity index = $(w - PL)/PI$ yields an indication of how stiff the soil is in its natural state. LI = 1 at LL; LI < 1 more plastic and more shear strength.

FI = flow index = slope of the flow line.

TI = toughness index = PI/FI.

A = activity = PI/(% passing 0.002 mm). A large A denotes the presence of active clay minerals such as montmorillonite in contrast to illite or kaolinite, the greater the swelling and shrinkage.

Terms Associated with Unit Weight, Composition, and Compaction

e = void ratio = volume of voids/volume of solids = $V_v/V_s = n/(1-n)$.

G_s = specific gravity of soil solids = γ_s/γ_o where γ_o = density of water.

G_w = specific gravity of water = $\gamma_w/\gamma_o \approx 1$.

n = porosity = $V_v/V = e/(1+e)$.

S = degree of saturation = % void volume that is filled with water = V_w/V_v.

V = volume (subscript v = voids, s = solids, w = water).

w = water content = W_w/W_s.

γ = unit weight.

γ_b = buoyant unit weight = $\gamma_t - \gamma_w = \{G_s - 1 - e(1-S)\}/(1+e)\gamma_w$.

γ_b = buoyant unit weight for a saturated soil = $\gamma_t - \gamma_w = (G_s - 1)/(1+e)\gamma_w$.

γ_{dry} = dry unit weight = $\gamma_{sat}/(1+w)$.

γ_o = unit weight of water at 4°C = 1 g/cm^3 = 62.4 lb/ft^3.

γ_{sat} = saturated unit weight = $W_s(1+w)/V$.

γ_t = total unit weight = $(G + Se)/(1 + e)\gamma_w = (1 + w)/(1 + e)G_s\gamma_w$.
γ_w = unit weight of water.

For saturation curve where $S = 1$:

$e = G_s\gamma_o/\gamma_{dry} - 1$ for any S
$S = (wG_s/e)(\gamma_o/\gamma_w) \Rightarrow Se = (wG_s)(\gamma_o/\gamma_w)$
$S = wG_s/\{(G_s\gamma_o)/\gamma_{dry} - 1\}$
$\gamma_{dry} = G_s\gamma_w/(1 + wG_s) = G_s\gamma_w/(1 + e)$
$\gamma_{dry} = G_s\gamma_w/(1 + wG_s) = G_s\gamma_w/(1 + wG_s/S) = \gamma_t/(1 + w)$ for $S = 1$
$\gamma_{sat} = G_s(1 + w)\gamma_o/(1 + e) = G_s(1 + w)\gamma_o/\{1 + (wG_s\gamma_o/(S\gamma_w))\}$
$\gamma_{wet} = (wG_s\gamma_o + G_s\gamma_o)/(1 + e)$.

Terms Associated with Soil Stresses, Bearing Capacity, Strength, and Displacement

B = bulk modulus = $E/\{3(1 - 2v)\}$
c = cohesion
c' = effective cohesion
D = constrained modulus = $E(1 - v)/\{(1 + v)(1 - 2v)\}$
E = elastic modulus
E_s = secant elastic modulus
E_t = tangent elastic modulus
FS = factor of safety in allowable stress design
g = acceleration of gravity = $9.806\,654$ m/s^2 = 32.174 ft/s^2
G = shear modulus = $E/\{2(1 + v)\}$
v_c = compressional wave velocity = $(E/\rho)^{0.5} = \{E/(\gamma/g)\}^{0.5}$
v_s = shear velocity = $(G/\rho)^{0.5} = \{G/(\gamma/g)\}^{0.5}$

Vertical:

k_v = coefficient or modulus of vertical subgrade reaction as in $p = ky$
N value = penetration resistance of split spoon in SPT
N_c = bearing capacity coefficient related to punching shear failure
N_q = bearing capacity coefficient related to depth of embedment of foundation
N_γ = bearing capacity coefficient related to width of foundation
P = pressure
P = load or force
q_a = allowable bearing capacity
q_u = ultimate bearing capacity
Q = load or force
SPT = Standard Penetration Test
y = deflection or displacement.

Lateral (horizontal):

k_h = constant of horizontal subgrade reaction
K_a = active earth pressure coefficient
K_o = coefficient of earth pressure at rest
K_p = passive earth pressure coefficient
α, β, i = angle
δ = deflection
ε = strain
ϕ = friction angle for soil
ϕ' = effective friction angle for soil
v = Poisson's ratio
σ = normal stress
τ = shear stress.

Terms Associated with Permeability and Flow of Water through Soil

A = cross-sectional area of flow
h = total hydraulic head
i = hydraulic gradient = $\Delta h / \Delta L$
k = permeability coefficient
L = length
q = flow rate = Q/t
Q = quantity of flow
t = time
v = flow velocity = Q/A.

Darcy's law governing one-dimensional water flow through saturated media:

$$Q = kiAt$$

Bernoulli energy equation for steady incompressible flow:

$$h = \text{total hydraulic head} = \text{constant}$$

$$= h_{elevation} + h_{pressure} + h_{velocity} = z + p/\gamma + v^2/(2g)$$

Laplace equation governing two-dimensional laminar flow through a saturated soil:

$$k_x^2 h/x^2 + k_y^2 h/y^2 = 0$$

Terms Associated with Soil Consolidation

a_v = coefficient of compressibility = $-de/dp$
c_v = coefficient of consolidation = $k(1+e)/(a_v \gamma_w)$
C_c = compression index = $(e - e_o)/\log_{10}(p/p_o)$
H = shortest drainage path
T = time factor for consolidation = $c_v t/H^2$
U = consolidation ratio.

Terzaghi's consolidation equation:

$$c_v^2 u/z^2 = u/t$$

References

AASHTO *Standard Specifications for Highway Materials and Methods of Sampling and Testing, Part II*, American Association of State Highway and Transportation Officials.

ASTM *Book of ASTM Standards*, American Society for Testing and Materials.

ASTM Committee D-18 *Procedures for Testing Soils*, American Society for Testing and Materials.

Bowles, J.E. (1970) *Engineering Properties of Soils and Their Measurement*, McGraw-Hill.

Germaine, J.T. and Gemaine, A.V. (2009) *Geotechnical Laboratory Measurements for Engineers*, John Wiley & Sons, Inc.

Lambe, T.W. (1951) *Soil Testing for Engineers*, John Wiley & Sons, Inc.

Lambe, T.W. and Whitman, R.V. (1969) *Soil Mechanics*, John Wiley & Sons, Inc.

Portland Cement Association (1972) *PCA Soil Primer*, Portland Cement Association, Skokie, IL.

Stevens, J. (1982) Unified Soil Classification System, *Civil Engineering – ASCE*, **52** (12), 61–62.

Taylor, D.W. (1948) *Fundamentals of Soil Mechanics*, John Wiley & Sons, Inc.

10

Soil Strength Properties

10.1 Discrete and Elastic Finite Element Models

A model of an elastic system requires that the soil strength properties of the soil material be characterized. There are two types of models used to represent the elastic body comprising a soil material. The model of an elastic system is one that breaks up the soil media into many finitely dimensioned elements. The strength of these elements is defined by the elastic stress–strain relation or E, the modulus of elasticity of the soil elements comprising the system. The other model employs springs to represent the entire soil media as a series of these springs. For this model a relation that relates load or stress to the displacement of the media is needed. The coefficient of subgrade reaction, k, is just such a relation between stress and displacement. These strength parameters along with others used in analytical expressions to describe the strength will be discussed in this chapter.

10.2 General Elasticity Equations Relating Stress and Strain

In any soil–structure system that is divided into a finite number of elements for analysis, the constitutive equations that express the stress–strain relations for the now elemental materials are at the heart of the numerical analysis.

For the general three-dimensional case of a linearly elastic material the following equations express the relation of normal or shear strain, ε or γ, to normal and shear stress, σ or τ:

$$\varepsilon_x = (1/E)(\sigma_x - \nu\sigma_y - \nu\sigma_z)$$

$$\varepsilon_y = (1/E)(\sigma_y - \nu\sigma_x - \nu\sigma_z)$$

$$\varepsilon_z = (1/E)(\sigma_z - \nu\sigma_x - \nu\sigma_y)$$

Solutions for Soil and Structural Systems using Excel and VBA Programs, First Edition. Robert L. Sogge.
© 2012 John Wiley & Sons, Ltd. Published 2012 by John Wiley & Sons, Ltd.

$$\gamma_{xy} = \tau_{xy}/G$$

$$\gamma_{yz} = \tau_{yz}/G$$

$$\gamma_{zx} = \tau_{zx}/G$$

where E is the modulus of elasticity, v is Poisson's ratio, and $G = E/\{2(1+v)\}$.

Alternatively, the elasticity equations could be written in terms of principal stresses and strains. The above elasticity equations could be presented as the following matrix of stress in terms of strain:

$$
\begin{bmatrix} \sigma_x \\ \sigma_y \\ \sigma_z \\ \tau_{xy} \\ \tau_{yz} \\ \tau_{zx} \end{bmatrix} = E/\{(1+v)(1-2v)\}
$$

$$
\times \begin{bmatrix} 1-v & v & v & & & \\ v & 1-v & v & & & \\ v & v & 1-v & & & \\ & & & 1-2v/2 & & \\ & & & & 1-2v/2 & \\ & & & & & 1-2v/2 \end{bmatrix} \begin{bmatrix} \varepsilon_x \\ \varepsilon_y \\ \varepsilon_z \\ \gamma_{xy} \\ \gamma_{yz} \\ \gamma_{zx} \end{bmatrix}
$$

10.2.1 Alternative Constitutive Equation Formulation

In a plane-strain formulation the values of v cannot equal 0.5 or the denominator in the elastic equation is zero. Thus, an undrained condition and the case where v exceeds 0.5 as failure is approached cannot be represented. This difficulty can be circumvented by writing the constitutive equations in the form (Clough and Woodward, 1967)

$$
\begin{bmatrix} \sigma_x \\ \sigma_y \\ \tau_{xy} \end{bmatrix} = \begin{bmatrix} B+G & B-G & 0 \\ B-G & B+G & 0 \\ 0 & 0 & G \end{bmatrix} \begin{bmatrix} \varepsilon_x \\ \varepsilon_y \\ \gamma_{xy} \end{bmatrix}
$$

where B = bulk modulus = $E/\{2(1+v)(1-2v)\}$ and G = shear modulus = $E/\{2(1+v)\}$.

If the bulk modulus is defined as $B = E/\{3(1-2v)\}$, the constitutive relations become (Hoeg, Christian, and Whitman, 1968)

$$
\begin{bmatrix} \sigma_x \\ \sigma_y \\ \tau_{xy} \end{bmatrix} = \begin{bmatrix} (3B+4G)/3 & (3B-2G)/3 & 0 \\ (3B-2G)/3 & (3B+4G)/3 & 0 \\ 0 & 0 & G \end{bmatrix} \begin{bmatrix} \varepsilon_x \\ \varepsilon_y \\ \varepsilon_{xy} \end{bmatrix}
$$

Post-failure behavior can be simulated by reducing the value of G, and even using a value of v greater than 0.5 if necessary, but maintaining the bulk modulus at its pre-failure level with $v < 0.5$. Thus, elements can sustain additional bulk stress after failure. Since the material will always contract under an all-around pressure, the bulk modulus cannot have $v > 0.5$. In order to simulate the undrained behavior which occurs during the loading of clays, it is desirable to have a theoretically infinite bulk modulus. This situation can be obtained by using a value of $v = 0.47$ (Huang, 1969).

10.2.2 Two-Dimensional Plane-Stress and Plane-Strain Constitutive Equations

As most analyses performed are for a two-dimensional idealization of a three-dimensional body, these equations will be written for such a case. The three-dimensional equations of elasticity can be reduced to two dimensions for the two-dimensional plane-strain and plane-stress states. For a two-dimensional plane-strain condition for which $\varepsilon_z = 0$, $\sigma_z = v(\sigma_x + \sigma_y)$, $\tau_{yz} = \tau_{zx} = 0$, the plane-strain constitutive equations in terms of stress are

$$\begin{bmatrix} \sigma_x \\ \sigma_y \\ \tau_{xy} \end{bmatrix} = E/\{(1+v)(1-2v)\} \begin{bmatrix} 1-v & v & 0 \\ v & 1-v & 0 \\ 0 & 0 & (1-2v)/2 \end{bmatrix} \begin{bmatrix} \varepsilon_x \\ \varepsilon_y \\ \gamma_{xy} \end{bmatrix}$$

For a plane-stress condition for which $\sigma_z = 0$, $\varepsilon_z = -v/(1-v)(\varepsilon_x + \varepsilon_y)$, $\tau_{yz} = \tau_{zx} = 0$, the constitutive equations are

$$\begin{bmatrix} \sigma_x \\ \sigma_y \\ \tau_{xy} \end{bmatrix} = E(1-v^2) \begin{bmatrix} 1 & v & 0 \\ v & 1 & 0 \\ 0 & 0 & (1-v)/2 \end{bmatrix} \begin{bmatrix} \varepsilon_x \\ \varepsilon_y \\ \gamma_{xy} \end{bmatrix}$$

The appropriate two-dimensional constitutive equations should be used for the analysis that is being conducted. Depending on the constitutive equations present in the analysis program, the E and v values either can be used directly or require modification. Plane-stress constitutive equations can be used for a plane-strain analysis by substituting the values $v = v/(1-v)$ and $E^* = E/(1-v^2)$ (v in this equation is the original v not v^*) for the v and E in the plane-stress equation formulation. Plane-strain equations can be used for a plane-stress analysis by substituting a value of $v^* = v/(1+v)$ and $E^* = E(1+2v)/(1+v)^2$ for the v and E input values to be used in an analysis based on a plane-strain equation formulation. In these conversions the value of the shear modulus, $G = E/\{2(1+v)\}$ does not change.

As long as the definition of E and the test conditions are the same, the value of E should not change. The values of E and v obtained from a test that models plane-stress or plane-strain state conditions would be different. A plane-stress state is one in which

$\sigma_z = 0$. A plane-strain state is one in which $\varepsilon_z = 0$ but $\sigma_z \neq 0$. Since $\sigma_z \neq 0$ in the plane-strain test, the modulus measured under plane-strain conditions will be greater than the one measured in a plane-stress state test. The elastic modulus determined from a confined compression test, D, can be related to the elastic modulus, E, by using the following relation:

$$E = D(1 + v)(1 - 2v)/(1 - v)$$

For $v = 1/3 \Rightarrow E = (2/3)D$.

10.3 Modulus of Elasticity and Poisson's Ratio

The modulus of elasticity or Young's modulus of a material can be expressed from the elasticity equations as

$$E = (\sigma_x - v\sigma_y - v\sigma_z)/\varepsilon_x$$

The modulus of elasticity, E, is commonly obtained from a uniaxial test on a cylinder having proper length to diameter proportions. E is defined as $E = \sigma_z/\varepsilon_z$ where, for a uniaxial test on a cylinder, $\sigma_x = \sigma_y = 0, \varepsilon_x = \varepsilon_y$ for the unconfined case, σ_z is the axial stress, and ε_z is the axial strain. From this test, Poisson's ratio is defined as $v = -\varepsilon_x/\varepsilon_z$ where ε_x is the lateral strain and ε_z is the vertical strain in the direction of the applied stress. For a material experiencing no volume change, $v = 0.5$. A dilatant material, such as a dense sand, would have a Poisson's ratio greater than 0.5.

As has been shown in Section 10.2, in a statically determinate system only the equilibrium equations are needed to solve for the forces in the system; the stresses are independent of E. In a statically indeterminate structure the values of E and v in the force–deformation equations are needed for computing the forces in the structure. In such a structure, the path that the load takes to the supports and the forces in the structure are dependent on the distribution of E throughout the structure, but not on the magnitude of E.

10.3.1 The Stress–Strain Curve

The curve relating stress to strain for a material may be linear elastic or inelastic. Also, it may be nonlinearly inelastic. A nonlinear elastic material does not exist since the inelastic stored internal energy cannot be recovered. These states are shown in Figure 10.1. Generally, we are concerned with loading states only, so rebound is not important, unless during the load application some soil regions are unloading.

The stress–strain curve is nonlinear and its elastic modulus and failure strength are dependent on the density or consistency of the soil as shown in Figure 10.2. In soil mechanics stress–strain curves are often plotted as the deviatoric stress, $(\sigma_1 - \sigma_3)$, to the principal strain, ε_1, or sometimes, to better define the failure point, σ_1/σ_3 versus ε_1. The variation of strength with confining pressure for such a plot is shown in Figure 10.3a.

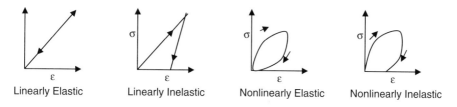

Figure 10.1   Material stress–strain curves.

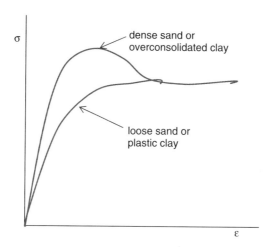

Figure 10.2   Stress–strain for soil.

The lack of linear elasticity can be neglected if the modulus is defined for a portion of the entire load range. Therefore, the definition of E requires the selection of some strain or stress level. A secant or tangent modulus value can fit such a description for a load increment as shown in Figure 10.4.

A secant modulus is defined as the slope of a chord between the origin and some other point at a chosen stress or strain level on the stress–strain curve. In general, this secant modulus is defined for stress levels at one-third to one-half the peak stress levels. The equation $E = (\sigma_1 - \sigma_3)/\varepsilon_1$ represents a secant modulus.

A tangent modulus is derived from the slope of a line tangent to the curve at some point. If a tangential representation of E is desired for use in an incremental analysis, then $E_t = d(\sigma_z - 2\nu\sigma_x)/d(\varepsilon_z)$. In a triaxial test where $\sigma_x = \sigma_y = $ constant throughout the test, $E_t = d\sigma_z/d\varepsilon_z$. In soil mechanics we use $E_t = d(\sigma_1 - \sigma_3)/d\varepsilon_1$. Such a definition will work in the constitutive equations since σ_3 is constant. The slope of a $(\sigma_z - \sigma_x)$ versus ε_z curve, a σ_z versus ε_z curve, or a $(\sigma_z - 2\nu\sigma_x)$ versus ε_z curve yields the same value for E_t. For a linear material both the secant and tangent moduli are the same. Typical values of E assuming a linear relation to one-third of the peak stress difference are

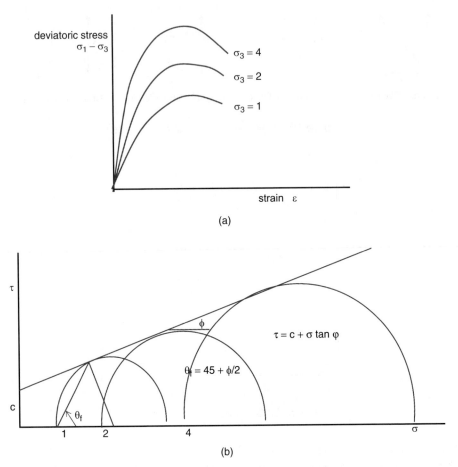

Figure 10.3 (a) Stress–strain for soil with varying confining pressure. (b) Mohr–Coulomb failure strength criteria for soil.

given in Table 10.1 along with v values and a correlation to ϕ values. The workbook **Soil Material Property Table** provides this table in the worksheet Matl-Prop.

10.3.2 Failure Strength Related to Confining Pressure Dependency

Different failure criteria apply to different types of materials. A failure state is governed by the failure law that is used to define it. Examples are the von Mises or distortion energy criterion, the Tresca or maximum shear stress criterion, and the Mohr–Coulomb failure criterion. Failure or yield laws are used to extrapolate the yield conditions of a simple uniaxial or triaxial test in order to determine the conditions of initial yielding when any or all of the stress components are present in a general stress state. The

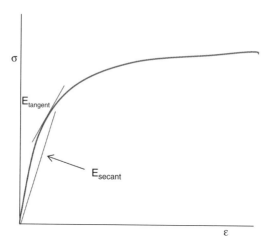

Figure 10.4 E_{tan} and E_{sec}.

criteria for continued yielding are described by equivalent stresses and flow rules. The equivalent stress is a measure of when the stress state has reached some point on the defined yield surface. Any difference in strength would be due to the different stress states causing failure to occur at different points on some assumed yield surface.

Yielding is defined as the onset of plastic (inelastic) deformations or the upper limit of elastic deformation. Fracture is the occurrence of distinct separated surfaces without measurable yielding. Types of fracture include brittle, shear, rupture, and ductile. There are failure theories for both plastic yielding and fracture failures.

For soil the Mohr–Coulomb failure criterion is commonly employed. The Mohr–Coulomb failure strength criterion is shown in Figure 10.3b. By this criterion the failure strength varies with confining pressure according to the relation

$$(\sigma_1 - \sigma_3)_f = (2c \cos \phi + 2\sigma_3 \sin \phi)/(1 - \sin \phi)$$

or for $c = 0$

$$(\sigma_1 - \sigma_3)_f = (\sigma_1 + \sigma_3) \sin \phi$$

This confining pressure dependency is a result of the frictional nature of soil. The angle of inclination of the failure plane to the major principal plane is $\theta_f = 45° + \phi/2$.

10.3.3 Elastic Modulus – Relation to Pore Water Pressure and Water Content

The value of E is dependent on the pore water pressure. Undrained and drained tests give different results. The relation between the two moduli can be developed considering the undrained portion of a consolidated–undrained (CU) test and the pore

Table 10.1 Soil strength relations.

Soil strength relations

Sands

N SPT Blows	Relative density	ϕ (deg)	Density (pcf)	k_{11}^* Vertical (kcf)[1]	k_1^{**} Lateral (kcf)[1]	E (ksf)	Poiss Ratio ν
<4	Very loose	<25	<90	<40	7	150 200	0.20
4 10	Loose	25 30	90	40 100	15	200 500	–
10 30	Medium dense	30 34	100	120 115	30	500 1000	0.36
30 50	Dense	34 38	115	600 122	60	1000 2000	–
>50	Very dense	>38	>122	>2000	120	2000 4000	0.40

(Vertical right-hand values: 120, 600, 2000)

Sand

[1] For submerged conditions use 1/2 of this value

$^*k_v = k_{11}^*\{(B+1)/(2B)\}^2$

$k_v = 1.227E/(1-n^2)$ for a rigid 1 ft × 1 ft plate =

$k_v = 1.05E/(1-n^2)$ for a flexible 1 ft × 1 ft plate

$^{**}k_h = k_1^* Z/B$ 1.35   1.15

E (ksf) = 40 N (approx)

1 ksf = 6.94 psi

1 kcf = 0.58 pci

Clay

N SPT Blows	Consistency	q_u (ksf)	k_{11}^* Vertical or Lateral (kcf)	E (ksf)	Poiss. Ratio ν
<2	Very soft	<0.5	<25	10 50	–
2 4	Soft	0.5	50	50 100	0.50
4 8	Firm	1	150	100 300	–
8 16	Stiff	2	350	300 700	–
16 32	Very stiff	4	500	700 1000	0.40
>32	Hard	>8	>500	1000 2000	–

Clay:

$k = k_{11}/B$

$c = q_u/2$

$q_u = N/4$

$^{**}k_{11}^* = 60 q_u = 15 N$

water parameter for an isotropic mineral skeleton. The result is

$$E = 3/\{2(1+v)\}E'$$

For $v = 0.3$ the undrained modulus is theoretically 1.15 times the effective modulus. In practice this ratio can go up to 3 or 4. Since the shear modulus $G = \tau_{xy}/\gamma_{xy} = (\sigma_1 - \sigma_2)/2/\gamma_{xy}$, it is unaffected by pore water pressure and the undrained and drained values are identical for the same effective stress. If the elastic modulus were defined from a $(\sigma_1 - \sigma_3)$ versus ε_1 plot, there would also be no difference between drained and undrained E values. The undrained strength of a clay soil is dependent on the initial confining pressure but not on any change from the initial value. All changes in confining pressure in an unconsolidated–undrained (UU) test would go into pore water pressure changes.

The modulus of elasticity of clays is relatively constant with depth but varies greatly with the water content of the soil, w. The variation with water content can be represented by the relation (Barkan, 1962) $E_2/E_1 = 1 - (w_2^2/w_1^2)$.

10.3.4 Elastic Modulus for Repeated Loading

Repeated or cyclic loading on granular materials results in a $\sigma - \varepsilon$ curves that are similar to those for unload–reload on the material. On unloading, the system behaves non-conservatively and inelastically with respect to the initial loading curve. The slopes of the unloading and reloading curve are approximated by the initial modulus as shown in Figure 10.5. Typically, the relation of the unloading–reloading modulus to the initial modulus varies from 1.1 to 3 for dense and loose soil, respectively. In general, for granular materials the second cycle of laboratory loading gives the best in-situ soil modulus by having the effect of sample disturbance balanced by the initial loading

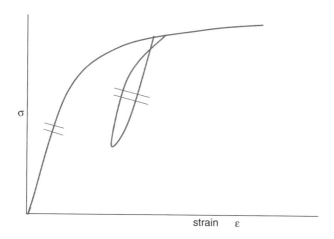

Figure 10.5 Unload–reload for plastic material.

cycle. Unlike granular soils, the modulus of clays is reduced by cyclic loading (Thiers and Seed, 1968).

10.3.5 Elastic Modulus for Dynamic Loading

The modulus of elasticity E of a material can be related to its compressional or P-wave velocity by the relation

$$V_p = (E/\rho)^{0.5} = \{E/(\gamma/g)\}^{0.5}$$

where g is the gravitational acceleration, γ is the unit weight, and ρ is the mass density of the material.

A similar relation exists between the shear wave (S-wave) velocity and the shear modulus.

The expression shown gives the elastic modulus for a rod. Such a velocity can be related to the elastic modulus for an infinite medium through the following Poisson's ratio relation:

$$E_{medium} = E_{rod}(1 + v)(1 - 2v)/\{2(1 - v)\}$$

Values of the compression wave velocity from which an elastic modulus can be calculated are presented in Table 10.2.

The values of an elastic modulus computed from a compression or shear wave velocity are considerably higher than the value obtained from the initial loading cycle of a static test. This factor can be on the order of 2–3.

10.3.6 Analytical Expressions for Elastic Modulus

10.3.6.1 Soil

Many researchers have expressed the failure characteristics of soil in terms of the confining stress invariant, $(\sigma_1 + \sigma_2 + \sigma_3)$, and the deviatoric stress $(\sigma_1 - \sigma_2)$, namely, Newmark (1960), Huang (1968), Girijavallabhan and Reese (1968), and Domaschuk and Wade (1969). A good nonlinear explicit representation of soil parameters E and v has been presented by Duncan and Chang (1970) and by Kulhawy, Duncan, and Seed (1969). Rather than use plastic stress–strain relations and flow rules, their approach assumes that a three-dimensional stress state fails in the same manner as the state described by the stress–strain curve for a triaxial test. Nonlinearity is incorporated using the hyperbolic relation suggested by Kondner and Zelasko (1963), $\sigma = \varepsilon/(a + b\varepsilon)$. The equation takes the form

$$\sigma_1 - \sigma_3 = \varepsilon/\{1/E_i + \varepsilon/(\sigma_1 - \sigma_3)_u\}$$

where $(\sigma_1 - \sigma_3)_u$ = the ultimate failure stress.

The confining pressure dependency of the initial tangent modulus, E_i, is represented by

$$E_i = Kp_{atm}(\sigma_3/p_{atm})^n$$

Table 10.2 Seismic velocity – dynamic elastic modulus relationship.

Material	Seismic velocities (comp, P-wave, V_p) – dynamic moduli					
	V_p (ft/s)		v	γ (pcf)	E (ksf)	E (ksi)
Near surface	800	2600	–	100	8975	62
Loess	1000	2000	–	100	6988	49
Gravel – loose	2000	3000	–	105	20380	142
Gravel – dense	3000	8000	–	124	116491	809
Clay	5000	8000	–	107	140396	975
Claystone	6000	11000	–	–	–	–
Sandstone	5000	15000	–	–	–	–
Limestone	10000	20000	0.28	135	943323	6551
Dolomite	16000	20000	0.30	140	1408696	9783
Salt	14000	21000	0.25	120	1141304	7926
Granite	13000	19000	0.25	145	1152795	8006
Gabbro	22000	24000	0.29	–	–	–
Dunite	26000	28000	0.26	–	–	–
Basalt	16000	21000	–	130	1381755	9596
Gneiss	10000	18000	0.18	–	–	–
Quartzite	16000	20000	0.13	–	–	–
Schist	12000	18000	0.18	130	908385	6308
Air	1130 (770 mph)					
Water – fresh	4700					
Above E is for a rod						
For E of infinite medium	$E = V_p^2 \, (\gamma/g)(1 + v)(1 - 2v)/\{2(1 - v)\}$					

where K is a dimensionless soil modulus number and n is a constant varying from 0.4 to 0.9. Both K and n are obtained from a log–log plot of E_i/p_{atm} versus σ^3/p_{atm}. Table 10.3 presents values of these parameters for this equation. The atmospheric pressure, p_{atm}, is given in the units desired for E.

The two-dimensional Mohr–Coulomb failure criterion whose parameters are determined in a three-dimensional test is used to predict the general three-dimensional failure state. The Mohr–Coulomb failure criterion is incorporated to determine the failure stress using

$$(\sigma_1 - \sigma_3)_f = (2c \cos \phi + 2\sigma_3 \sin \phi)/(1 - \sin \phi)$$

and R_f, a failure ratio, typically 0.9, that defines $(\sigma_1 - \sigma_3)_f = R_f(\sigma_1 - \sigma_3)_u$.

By differentiating the hyperbolic equation, the tangent modulus, E_t, is derived as follows:

$$E_t = (1/E_i)/\{1/E_i + \varepsilon/(\sigma_1 - \sigma_3)_u\}^2$$

As the developers of this derivation state, since the reference state for strain is completely arbitrary and because stresses can be calculated more accurately than strains in many soil mechanics problems, this equation expressed in terms of stress

Table 10.3 Duncan relation for E_i.

		$E_i = Kp_a(\sigma_3/p_a)^n$		
E_i (ksf)	K	p_{atm} (ksf)	σ_3 (ksf)	n
845	500	2.12	1	0.30
845	500	14.7	1.0	0.30

Soil group	ϕ (deg)		K	n	E_i (ksf)	v_i	E_i (ksf)
	Low σ_3	High σ_3					
GW	47	35	500	0.3	800	0.32	845
GP	46	38	1800	0.3	3000	0.38	3043
SW	50	35	300	0.5	500	0.3	436
SP	40	30	1200	0.5	1800	0.54	1746

Note: The initial modulus of elasticity, E_i, has been computed based on a soil density of 130 pcf (relative density = 100%) and a confining pressure of 1 ksf or approximately 8 ft of overburden.

difference is preferable. Using

$$\varepsilon = \{(\sigma_1 - \sigma_3)/E_i\}/\{1/E_i + e/(\sigma_1 - \sigma_3)_u\}$$

yields

$$E_t = E_i/\{1 - (\sigma_1 - \sigma_3)/(\sigma_1 - \sigma_3)_u\}^2$$

A secant version of the modulus, E_{sec}, could be developed in a manner similar to what Richard did as shown in Section 10.3.6.3. The Duncan formulation was modified and put in terms of a tangent elastic modulus and a bulk modulus (Duncan et al., 1980). Further related modifications were made by Selig (1988).

10.3.6.2 Concrete

A good analytical representation of the f_c versus σ failure curve for concrete can be readily made by using a second-order polynomial, a parabola, to describe the non-dimensionalized $\sigma - \varepsilon$ equation. It then takes the form

$$(f_c/f_c') = a(\varepsilon/\varepsilon_f)^2 + b(\varepsilon/\varepsilon_f) + c$$

where ε_f is the concrete strain at failure. Substituting the boundary conditions

$$f_c/f_c' = 1 \text{ at } \varepsilon/\varepsilon_f = 1$$
$$f_c/f_c' = 0 \text{ at } \varepsilon/\varepsilon_f = 0$$
$$d(f_c/f_c')/d(\varepsilon/\varepsilon_f) = 0 \text{ at } \varepsilon/\varepsilon_f = 1$$

yields the constants $a = -1$, $b = 2$, and $c = 0$ and the equation

$$(f_c/f_c') = -(\varepsilon/\varepsilon_f)^2 + 2(\varepsilon/\varepsilon_f)$$

Concrete typically fails at a strain, ε_f, of 0.002. The failure ε based on the initial slope of the parabola defining the failure curve is $\frac{1}{2}\varepsilon_f$ or 0.001. For this curve $E = f_c'/(\varepsilon_f/2)$. For 3000 psi concrete, $E_i = 57000(f_c')^{0.5} = 3122$ ksi, which is very close to the value of $f_c'/(\varepsilon_f/2)$. Similarly, the failure strain based on the concrete modulus E is 0.00096. The similar values show how well this equation describes the failure curve. This parabolic stress–strain curve for concrete is presented in the worksheet Stress-Strain Prop of the workbook **Reinf Concrete** presented in Chapter 8.

Furthermore, computing the f_c/f_c' ratio when the concrete cylinder reaches $e_u = 0.003$, the maximum usable or ultimate failure strain, yields a ratio of 0.75, which is a reasonable description of the curve:

$$E_t = df_c/d\varepsilon = E(1 - \varepsilon/\varepsilon_f)$$
$$E_{sec} = (E/2)(2 - \varepsilon/\varepsilon_f)$$

It is convenient to put these equations in terms of stress rather than strain since stresses can often be calculated to greater accuracy than strains. Using the original equation in the form

$$\varepsilon/\varepsilon_f = 1 \pm (1 - f_c/f_c')^{0.5}$$

yields

$$E_t = \pm E(1 - f_c/f_c')^{0.5} \quad \text{and} \quad E_{sec} = E/2\{1 \pm (1 - f_c/f_c')^{0.5}\}$$

10.3.6.3 Steel

Steel has a constant stress difference at failure that is governed by the maximum shear stress criterion, for which $(\sigma_1 - \sigma_3)_f$ equals the yield stress from a uniaxial test. The failure strength of a steel material in tension or compression is essentially the same as shown in Figure 10.6a. The failure angle is approximately 45° since ϕ is approximately zero as shown in Figure 10.6b. This criterion is applicable to the plastic nature that steel displays.

A nonlinear stress–strain curve formulation applicable to steel and many other materials was published by Richard and Blacklock (1969). It uses the same hyperbolic equation on which Duncan's work is based but employs a parameter, n, different than 1 that permits changes in the $(\sigma_1 - \sigma_3/(\sigma_1 - \sigma_3)_u$ ratio different than 0.5:

$$\sigma = E\varepsilon/\{1 + (E\varepsilon/\sigma_o)^n\}^{1/n} \text{ where } E = \sigma_o/\varepsilon_o$$

or in non-dimensional form:

$$(\sigma/\sigma_o) = (\varepsilon/\varepsilon_o)/\{1 + (\varepsilon/\sigma_o)^n\}^{1/n}$$

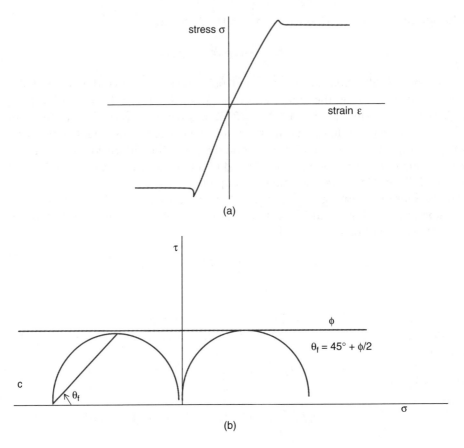

(a)

(b)

Figure 10.6   (a) Stress–strain for steel. (b) Mohr–Coulomb failure strength criteria for steel.

Using values of 10 or 20 for n permits the sharp steel failure curve to be modeled. By just using a normal hyperbolic equation (as in Duncan's formulation with parameter $n = 1$) the failure stress will not approach the true yield change of steel. The plot of σ/σ_o versus n (Figure 10.7b) for $\varepsilon/\varepsilon_o = 1$ gives the value of n required to develop the required stress ratio at this strain level.

As discussed in the section dealing with a soil hyperbolic stress–strain curve, it is desirable for many soil problems to express the above equations in terms of stress rather than strain.

Expressing the $\sigma - \varepsilon$ curve in terms of stress yields the equation

$$(\varepsilon/\varepsilon_o) = (\sigma/\sigma_o)/\{1 - (\sigma/\sigma_o)^n\}^{1/n}$$

The differential form of the constitutive relationship yields E_t in terms of strain as the inverse power equation

$$E_t/E = \{1 + (\varepsilon/\varepsilon_o)^n\}^{(n+1)/n}$$

and in terms of stress as

$$E_t/E = \{1 - (\sigma/\sigma_o)^n\}^{(n+1)/n}$$

The secant modulus is expressed in terms of strain and stress in the following equations:

$$E_{sec}/E = \{1 + (\varepsilon/\varepsilon_o)^n\}^{1/n} \quad \text{and} \quad E_{sec}/E = \{1 - (\sigma/\sigma_o)^n\}^{1/n}$$

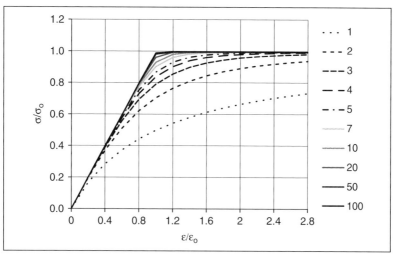

$$\sigma/\sigma_o = (\varepsilon/\varepsilon_o)/(1 + (\varepsilon/\varepsilon_o)^n)^{1/n}$$

(a)

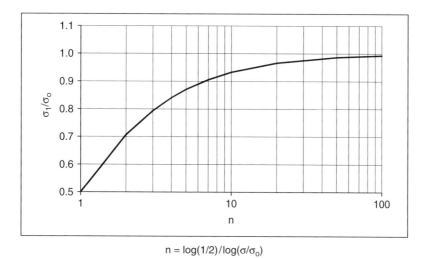

$$n = \log(1/2)/\log(\sigma/\sigma_o)$$

(b)

Figure 10.7 (a)–(g) Richard stress–strain curves for steel.

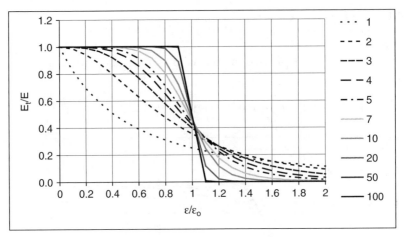

$$E_t/E = 1/(1+(\varepsilon/\varepsilon_o)^n)^{(n+1)/n}$$

(c)

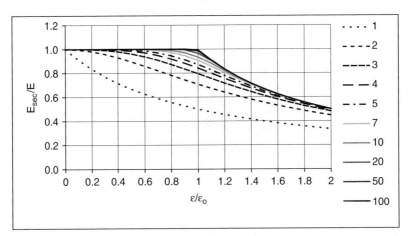

$$E_{sec}/E = 1/(1+(\varepsilon/\varepsilon_o)^n)^{1/n}$$

(d)

Figure 10.7   *(continued)*

Richard and Blacklock (1969) also developed relations for the variation of Poisson's ratio with stress level

$$\nu_t = 0.5 - (0.5 - \nu)(E_t/E)$$

and a ν_s value by substituting E_s for E_t in the same equation.

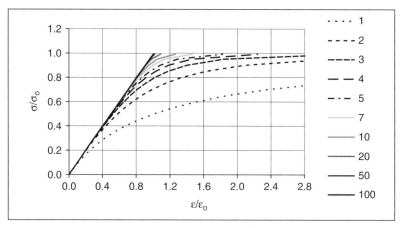

$$\varepsilon/\varepsilon_o = (\sigma/\sigma_o)/ (1-(\sigma/\sigma_o)^n)^{1/n}$$

(e)

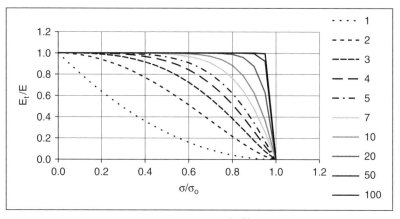

$$E_t/E = (1-(\sigma/ \sigma_o)^n)^{(n+1)/n}$$

(f)

Figure 10.7   (continued)

Non-dimensionally these relations are expressed as

$$\nu_t/\nu = E_t/E$$
$$\nu_s/\nu = E_s/E$$

Plots of these equations are given in Figures 10.7a–g. The workbook **Figure 10.7a – g
Richard-Stress-Strain Relations for Steel** also provides plots of these relations.

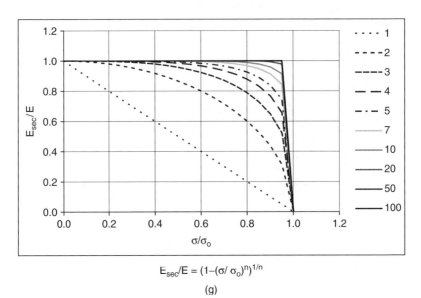

$$E_{sec}/E = (1-(\sigma/\sigma_o)^n)^{1/n}$$

(g)

Figure 10.7   *(continued)*

A generalized versatile elastic–plastic relation for stress–strain and their mathematical description for the various conditions of strain hardening, strain stiffening, strain softening, and a gap element have been presented by Richard and Abbott (1975) as shown in Figure 10.8a,b.

10.3.7 Secant and Tangent Modulus Values for Iterative and Incremental Analysis

There are two types of nonlinear analyses used for systems involving soils and structures, namely, iterative and incremental. An iterative analysis applies the total amount of the body forces and applied loads instantaneously and then iterates to the solution by varying the moduli within the system to be consistent with the stresses in the system. Numerical difficulties caused by instabilities in the iterations and lack of convergence can result with this type of analysis. Secant moduli are used in iterative types of analysis where the full load is applied and iteration toward the compatible load–deformation state is conducted.

Another type of analysis that applies the loads in increments is denoted as an incremental type of analysis. For soil problems where a modeling of the construction sequence consisting of backfill or cut steps is performed, tangent moduli are used in an incremental type of analysis. In this analysis increments of load are applied up to the full load. The tangent modulus used in an incremental analysis can be negative for soils strain softening beyond the peak point on the $\sigma-\varepsilon$ curve. Computationally,

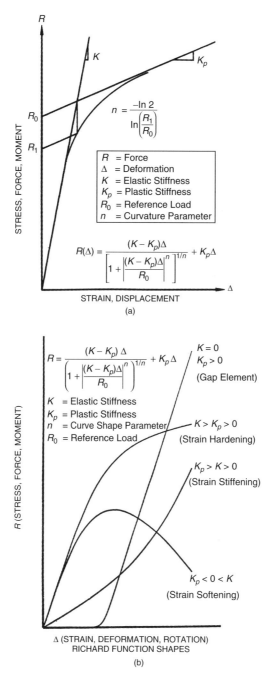

Figure 10.8 (a) Richard stress–strain curve for strain-hardening material. (b) Richard stress–strain curve for generalized material.

at this point two stress states exist and further loading can be carried out only by incrementing displacements. Moduli representations like Duncan's and Richard's provide a continually increasing stress–strain curve that avoids analytical instabilities with computations.

10.3.8 General and Local Failure Conditions

The conditions of local and general shear failure, as in bearing capacity, are related to the material's stress–strain curve. Loose sands display strain hardening or a plastic $\sigma - \varepsilon$ relation. The stress–strain curve of dense sands and overconsolidated clays display strain-softening behavior past the peak stress difference, as shown in Figure 10.2. Concrete displays this same behavior as was shown in Section 8.1 using a parabolic description of the stress–strain curve.

Loose soils show large movements as a local shear failure of the system is engaged. Punching shear is an example of such movement. Dense material behavior results in local shear failure as local zones experience large deformations that are great enough to result in failure, even though the ultimate strength for the system has not been reached. In the case of a dense sand, the load is transferred out of the failing region causing other areas to receive more load, until they reach the point where the entire general region begins failing and strain softening of the general system load–deformation curve occurs.

Materials having a strain-softening property result in systems displaying general and local failure. For a strain-softening system, as more elements framing into a node fail (strain soften), the nodal stiffness will become negative. Physically, this is a region of local failure. The local instability that the negative terms indicate does not mean that general failure is occurring. As long as the load–deformation relation for the entire system is strain hardening, the entire system remains stable. Arching or bridging across and around the zones of local instability occurs as the load is transferred out of the unstable region.

10.3.9 The Relation between v, ϕ, and K_o

Poisson's ratio can be related to the coefficient of earth pressure at rest, K_o, by substituting $\sigma_2 = \sigma_3 = K_o\sigma_1$ in the elastic equation

$$\varepsilon_2 = (1/E)(\sigma_2 - v\sigma_1 - v\sigma_3)$$

For no movement in the lateral direction $\varepsilon_2 = 0$, thus $\sigma_2 = v\sigma_1 + v\sigma_3$ yielding $v = K_o/(1 + K_o)$ or, inversely,

$$K_o = v/(1 - v)$$

An expression for K_o, the at-rest earth pressure coefficient, given by Jaky for a cohesionless material, is $K_o = 1 - \sin\phi$. Equating Jaky's expression for K_o with the expression for K_o in terms of v yields the relation

$$v = (1 - \sin\phi)/(2 - \sin\phi)$$

10.3.10 Analytical Representation of Poisson's Ratio

In a triaxial test, in which $E = (\sigma_z - 2\nu\sigma_x)/\varepsilon_z$, and $\nu\sigma_x = \nu\sigma_y$, the value of Poisson's ratio, ν, is constantly changing throughout the test, generally increasing with stress level. Initial values are usually less than 0.5, but approach 0.5 during shear failure when a positive volume change occurs. A value of 0.5 occurs in a state of no volume change.

Dense sand samples that dilate during shear, creating positive volume change, can have a ν greater than 0.5 (Taylor, 1948, p. 331).

Materials such as dense sands and overconsolidated clays display dilatant properties along with their strain-softening behavior. This behavior is described by ν being greater than 0.5. For loose sands and normally consolidated clays, n values greater than 0.5 only occur near failure. Instantaneous or tangent values of ν are often greater than 0.5 even though the total or secant ν to that point may be less than 0.5.

A theory for dilatant soils was proposed by Hermann (1965) and later modified by Hwang, Ho, and Wilson (1969). Smith and Kay (1971) dealt with a contractive or dilative strain-hardening soil. A general representation that can be used has been mentioned previously. By using a bulk and shear modulus to define the constitutive equations, all volume change can be attributed to shear deformation with normal stress deformation being negligible. Dilation is accounted for by permitting values of Poisson's ratio greater than 0.5 in the shear modulus.

A nonlinear expression for a tangent Poisson's ratio has been developed by Kulhawy, Duncan, and Seed (1969) in a manner consistent with their definition of tangent modulus. The relationship, which is confining-pressure dependent, has been modified and put in terms of E_t. Therefore, ν at the stress level for which E is desired should be used.

10.3.11 Typical E and ν Values

Typical values of E assuming a linear relation to one-third of the peak stress difference are given in Table 10.1 along with ν values and a correlation to ϕ values. As stated previously, the modulus of elasticity of clays is relatively constant with depth but varies greatly with the water content, w, according to the relation (Barkan, 1962)

$$E_2/E_1 = 1 - w_2^2/w_1^2$$

Values of K and n in the Duncan–Chang equation and initial ν_t are given in Table 10.3 for material having a relative density, D_r, of 100% in a drained triaxial test. Values of K can be converted to other densities by the relation $KN/D_r = $ constant. Table 4.4.7.2.2A in AASHTO (2002) provides correlations between elastic modulus and SPT N values. Samtani and Nowatzki (2006) provide corrections for N values based on overburden depth and hammer efficiency.

10.4 Coefficient of Subgrade Reaction

The coefficient of vertical subgrade reaction was originally developed by Terzaghi (1955). The coefficient of lateral or horizontal subgrade reaction has been developed by using three different approaches.

10.4.1 Terzaghi Relation for k_v and k_h

The coefficient (sometimes called modulus) of vertical subgrade reaction, k, relates pressure, p, to deflection, y, at the system level, $p = ky$. The units for k are force/length^3. There are different values depending on whether the load is vertically or horizontally applied.

Values of the coefficient of vertical subgrade reaction for a 1 ft × 1 ft square plate or a 1 ft wide strip are given in Table 10.1. For realistic foundation sizes, the influence of scale enters. For use with square footings on sand, the k of a 1 ft × 1 ft square plate must be multiplied by $\{(B+1)/(2B)\}^2$, where B is the footing width, according to Terzaghi (1955, p. 303). The value of k for use with square footing on clays must be multiplied by $1/B$. For a lateral loading situation, the coefficient of horizontal subgrade reaction is

$$k_h = l_h z/Dy$$

where

$\quad\quad$ D = height of structure
$\quad\quad$ z = the depth of the point where the pressure is being computed
$\quad\quad$ l_h = constant of horizontal subgrade reaction that is independent of scale.

Values of l_h given by Terzaghi (1955, p. 319) are summarized in Table 10.1. Examples of the use of these parameters are presented in Chapters 16–20.

10.4.2 Bowles Relation for k_h

A relation that has been developed by Bowles (1977) for k_h is of the form $k_h = a + bz$, where the coefficients a and b are related to the bearing capacity factors N_c, N_q, and N_γ (these bearing capacity factors will be employed in Chapter 13 to determine the bearing capacity of shallow and deep foundations):

$$(aa) = e^{(\pi \times 3/4 - \phi/2)\tan\phi}$$

$$N_q = (aa)^2/\{2\cos^2(45° + \phi/2)\}$$

$$N_\gamma = 1.8(N_q - 1)\tan\phi$$

$$N_c = (N_q - 1)/\tan\phi$$

$$k_h = a + bz^1 = 12(cN_c + 0.5\gamma BN_\gamma) + 12\gamma N_q z^1$$

Letting L be the maximum value of z and using a shape modifier $(z/L)^n$ yields

$$k_h = a + bL^1(z/L)^n = 12(cN_c + 0.5\gamma BN_\gamma) + 12\gamma N_q L^1(z/L)^n$$

Values for the coefficients a and b are presented in Table 10.4 for various values of ϕ.

Table 10.4 Relation between k_h and bearing capacity coefficients.

$$k_h = a + bz^1 = 12(cN_c + 0.5\gamma BN_\gamma) + 12\gamma N_q z^1 \qquad\qquad B = 1 \qquad\qquad \gamma$$
$$k_h = a + bL^1(z/L)^n = 12(cN_c + 0.5\gamma BN_\gamma) + 12\gamma N_q L^1(z/L)^n \qquad L = \max z \text{ distance} \quad 0.120$$

ϕ	c	(aa)	N_c	N_γ	N_q	a	B	n
26	0	2.83	27	12	14	8	20	1
28	0	3.07	32	16	18	12	26	1
30	0	3.35	37	22	22	16	32	1
33	0	3.83	48	37	32	26	46	1
38	0	4.86	77	85	62	61	89	1
40	0	5.39	96	121	81	87	117	1

10.4.3 E and k_h Developed from Lateral Wall Movement

The value of k_h can be developed from a value of E. A value for the value of E in terms of z, the distance below the ground surface, can be developed using a relation between wall movement toward and away from a fill. Chapter 12 will show this relation.

For the active case that is developed at a strain of approximately 0.5%,

$$E = \sigma_h/\varepsilon = (k_o - k_a)\gamma z/\varepsilon$$
$$E/z = (k_o - k_a)\gamma/\varepsilon$$

For the passive case that is developed at a strain of approximately 5%,

$$E = \sigma_h/\varepsilon = (k_p - k_o)\gamma z/\varepsilon$$
$$E/z = (k_p - k_o)\gamma/\varepsilon$$

These values are derived in Table 10.5 for various values of ϕ.

Table 10.5 Determination of E from wall movement.

ϕ	ν	Strain (%) = −0.50% K_a	0.00% $K_o = 1 - \sin\phi$	5.00% K_p	0.110 E/z Active	Passive	Coefficient of horizontal subgrade reaction $p = kx$ $k = E$
20	0.40	0.49	0.66	2.04	3.7	3.0	Elastic modulus
24	0.37	0.42	0.59	2.37	3.8	3.9	$E = \sigma/\varepsilon$
28	0.35	0.36	0.53	2.77	3.7	4.9	Active
30	0.33	0.33	0.50	3.00	3.7	5.5	$E = (k_o - k_a)\gamma z/\gamma$
32	0.32	0.31	0.47	3.25	3.6	6.1	$E/z = (k_o - k_a)\gamma/\gamma$
36	0.29	0.26	0.41	3.85	3.4	7.6	Passive
40	0.26	0.22	0.36	4.60	3.1	9.3	$E = (k_p - k_o)\gamma z/\varepsilon$
					Average = 3.6	5.8	$E/z = (k_p - k_o)\gamma/\varepsilon$

Knowing that k is approximately E, the expression $k_h = (3-10)z$ in kcf results. Such a value is in line with the k values shown in Table 10.1 for $k_h = kz$.

10.4.4 Relation between k_v and E

The relation between E and k_v for a soil is generally $k_v \approx E$ for the units of kips and ft. The relation between E and k_h used for laterally loaded piles and buried culverts is $k_h = k_1 z \approx E$, again for the units of kips and ft.

An explicit relationship between the modulus of elasticity and the coefficient of vertical subgrade reaction can be developed by using the theory of elasticity solution for the settlement below a rigid or flexible square plate of size 1 ft × 1 ft. The settlement, δ, for a footing having a least lateral dimension B subject to a contact pressure of intensity q is

$$\delta = qB(1 - v^2)/EI$$

where I is an influence factor which depends on the shape and rigidity of the footing and the depth of the consolidating layer. For a square rigid footing $I = 0.815$ and for a square flexible footing I has an average value of 0.95. Using the definition of k yields

$$k = q/\delta = 1/\{B(1 - v^2)I/E\}$$

Therefore $k = 1.227E/(1 - v^2)$ for a rigid 1 ft × 1 ft plate and $k = 1.05E/(1 - v^2)$ for a flexible 1 ft × 1 ft plate.

Using these conversions with typical E and v values presented in Table 10.1, the magnitude of k_v for sands ranging from loose to dense is 400–1400 kcf, while for stiff and hard clays it is 250 and 560 respectively.

10.5 Mathematical Descriptions of Curves Using Program Curve Fit

It is often beneficial to develop a relation between two parameters (such as ϕ and k_1) for automatic calculation of such parameters in the execution of a spreadsheet or program. Such a relationship can be developed by fitting a curve to the data points relating both parameters. As an example, for sands the values of k_h, k_v, and unit weight, γ, can be automatically calculated based on the input friction angle, ϕ, from the relations

$$k_h = 4.055 \exp\{(\ln(\phi) - 2.4444)^{2/0.4576}\}$$

$$k_v = 0.4109(\phi^{0.0562\phi})$$

$$\gamma(\text{pcf}) = 1/(0.03309 - 0.006823 \ln(\phi))$$

using workbook **CurveFit** with data files phi-k_h, phi-k_v, and phi-den consisting of 16 x-y data points in each file. These equations have been developed using the program CurveFit (Cox, 1992) converted to an Excel macro procedure. Specifically, **Example 10.1**

Table 10.6 Equations fitted with CurveFit.

Equation number	Equation	Curve
1	$Y = A + B^*X$	Linear
2	$Y = B^*X$	Linear through origin
3	$Y = 1/(A + B^*X)$	Reciprocal of linear
4	$Y = A + B^*X + C/X$	Combined linear and reciprocal
5	$Y = A + B/X$	Hyperbola
6	$Y = X/(A^*X + B)$	Reciprocal of hyperbola
7	$Y = A + B/X + C/(X^*X)$	Second order hyperbola
8	$Y = A + B^*X + C^*X^*X$	Parabola
9	$Y = A^*X + B^*X^*X$	Parabola through origin
10	$Y = A^*X \wedge B$	Power
11	$Y = A^*B^{\wedge}X$	Modified power
12	$Y = A^*B^{\wedge}(1/X)$	Root
13	$Y = A^*X^{\wedge}(B^*X)$	Super geometric
14	$Y = A^*X^{\wedge}(B/X)$	Modified geometric
15	$Y = A^*e^{\wedge}(B^*X)$	Exponential
16	$Y = A^*e^{\wedge}(B/X)$	Modified exponential
17	$Y = A + B^* \ln(X)$	Logarithmic
18	$Y = 1/(A + B^* \ln(X))$	Reciprocal of logarithmic
19	$Y = A^*B^{\wedge}X^*X^{\wedge}C$	Hoerl function
20	$Y = A^*B^{\wedge}(1/X)^*X^{\wedge}C$	Modified Hoerl function
21	$Y = A^*e^{\wedge}(((X-B)^{\wedge}2)/C)$	Normal distribution
22	$Y = A^*e^{\wedge}((\ln(X) - B)^{\wedge}2/C)$	Log-normal distribution
23	$Y = A^*X^{\wedge}B^*(1 - X)^{\wedge}C$	Beta distribution
24	$Y = A^*(X/B)^{\wedge}C^*e^{\wedge}(X/B)$	Gamma distribution
25	$Y = 1/(A^*(X+B)^{\wedge}2 + C)$	Cauchy distribution

(**phi-den**) shows how the equation above was developed. This **CurveFit** program provides more equations than those in the "add a trendline" feature of Excel charts. Table 10.6 provides the equation descriptions. Plotted results from using **CurveFit** with various bearing capacity and other soil parameters are presented in the workbook **CrvFitRelations**.

References

AASHTO (2002) *Standard Specifications for Highway Bridges*, 17th edn, American Association of State Highway and Transportation Officials, Last Update for Non-LRFD Design.

Barkan, D.D. (1962) *Dynamics of Bases and Foundations*, McGraw-Hill.

Bowles, J.E. (1977) *Foundation Analysis and Design*, McGraw-Hill.

Clough, R.W. and Woodward, R.J. III (1967) Analysis of embankment stresses and deformations. *Journal of the Soil Mechanics and Foundations Division, ASCE*, **93** (SM4), 529–549 (Proc Paper 5329).

Cox, T.S. (1992) Program CurveFit, QBasic ver 2.25a (5/9/92), Easley, SC (based on Kolb, W.M. (1984) *Curve Fitting for Programmable Calculators*, 3rd edn, SYNTEC, Bowie, MD).

Domaschuk, L. and Wade, N.H. (1969) A study of bulk and shear moduli of a sand. *Journal of the Soil Mechanics and Foundations Division, ASCE*, **95** (SM2), 561–581 (Proc Paper 6461).

Duncan, J.M., Byrne, P., Wong, K.S. et al. (1980) Strength, Stress-Strain and Bulk Modulus Parameters for Finite Element Analysis of Stresses and Movements in Soil Masses. Report No. UCB/GT/80-01, University of California, College of Engineering, Berkeley.

Duncan, J.M. and Chang, C.-Y. (1970) Nonlinear analysis of stress and strain in soils. *Journal of the Soil Mechanics and Foundations Division, ASCE*, **96** (SM5), 1629–1653 (Proc Paper 7513).

Girijavallabhan, C.V. and Reese, L.C. (1968) Finite element method applied to some problems in mechanics. *Journal of the Soil Mechanics and Foundations Division, ASCE*, **94** (SM2), 473–496 (Proc Paper 5864).

Hermann, L.R. (1965) Elasticity equations for incompressible and nearly incompressible materials by a variational theorem. *AIAA Journal*, **3**, 1896–1900.

Hoeg, K., Christian, J.T., and Whitman, R.V. (1968) Settlement of strip load on elastic-plastic soil. *Journal of the Soil Mechanics and Foundations Division, ASCE*, **94** (SM2), 431–443 (Proc Paper 5850).

Huang, Y.H. (1968) Stresses and displacements in nonlinear soil media. *Journal of the Soil Mechanics and Foundations Division, ASCE*, **94** (SM1), 1–19 (Proc Paper 5714).

Huang, Y.H. (1969) Finite element analysis of nonlinear soil media. ASCE Symposium on Application of Finite Element Methods in Civil Engineering, Vanderbilt University, Nashville, TN, November, pp. 663–690.

Hwang, C.T., Ho, M.K., and Wilson, N.E. (1969) Finite element analysis of soil deformations. ASCE Symposium on Application of Finite Element Methods in Civil Engineering, Vanderbilt University, Nashville, TN, November, pp. 729–746.

Kondner, R.L. and Zelasko, J.S. (1963) A hyperbolic stress-strain formulation for sands. Proceedings of the 2nd Pan American Conference on Soil Mechanics & Foundation Engineering, Brazil, vol. 1, pp. 289–324.

Kulhawy, F.H., Duncan, J.M., and Seed, H.B. (1969) Finite Element Analysis of Stresses and Movements in Embankments During Construction. Geotechnical Engineering Report No. TE-69-4, University of California, Berkeley.

Lambe, T.W. and Whitman, R.V. (1969) *Soil Mechanics*, John Wiley & Sons, Inc.

Lee, K.L. (1970) Comparison of plane strain and triaxial tests on sand. *Journal of the Soil Mechanics and Foundations Division, ASCE*, **96** (SM3), 901–923 (Proc Paper 7276).

Newmark, N.M. (1960) Failure hypothesis for soils. ASCE Research Conference on the Shear Strength of Cohesive Soil, University of Colorado, Boulder, June, pp. 17–32.

Richard, R.M. and Abbott, B.J. (1975) A versatile elastic-plastic stress-strain formula. *Journal of the Engineering Mechanics Division, ASCE*, **101** (EM4), 511–514 (Technical Note).

Richard, R.M. and Blacklock, J.R. (1969) Finite element analysis of inelastic structures. *AIAA Journal*, **7** (3), 432–438.

Samtani, N.C. and Nowatzki, E.A. (2006) Soils and Foundations, vols. I and II. Report No. FHWA-NHI-06-088 and -089, FHWA.

Selig, E.T. (1988) Soil parameters for design of buried pipelines. ASCE Conference on Pipeline Infrastructure, Boston, MA, June, pp. 99–116.

Smith, I.M. and Kay, S. (1971) Stress analysis of contractive or dilative soil. *Journal of the Soil Mechanics and Foundations Division, ASCE*, **97** (SM7), 981–997 (Proc Paper 8263).

Taylor, D.W. (1948) *Fundametals of Soil Mechanics*, John Wiley & Sons, Inc.

Terzaghi, K. (1955) Evaluation of coefficients of subgrade reaction. *Geotechnique*, **5** (4), 297–326.

Thiers, G.R. and Seed, H.B. (1968) Cyclic stress-strain characteristics of clay. *Journal of the Soil Mechanics and Foundations Division, ASCE*, **94** (SM2), 555–569 (Proc Paper 5871).

11

Stresses in an Elastic Half-Space

11.1 Closed-Form Elasticity Solutions

Elasticity solutions can be closed-form solutions to the formulation of equilibrium, force–deformation, and compatibility equations. These solutions are referred to as classical solutions. The configuration of some problems fits the assumptions that have to be made for this approach and can be applied. An example of a closed-form solution is the Burns and Richard (1964) solution presented in Chapter 20 for the stresses and displacements in an elastic circular cylindrical shell embedded in an elastic medium loaded by a surface overpressure.

11.2 Lateral Stresses against a Wall Restrained from Movement due to Point, Line, and Strip Loading

Modified theory of elasticity equations is used to compute the lateral pressure, σ_h, against a rigid wall of height H, at a depth of z ($= nH$), below the top surface of the wall, for a point Q (lb), line p (lb/ft), or strip of intensity, g, applied at the surface over a width w (lb/sq ft) load applied at a distance of $x = mH$ away from the face of the wall (Bowles, 1977, pp. 355–357; Spangler, 1936; AASHTO, 2010). Terzaghi (1954) found that pressures against a rigid retaining wall were approximately double those of the Boussinesq elastic equations.

11.3 Boussinesq Equation

This is used to solve for the vertical, σ_z, and lateral, σ_r, σ_θ, stresses beneath the surface of an elastic half-space (semi-infinite weightless mass) due to a vertical point load applied at the surface (Poulos and Davis, 1974, p. 16).

Solutions for Soil and Structural Systems using Excel and VBA Programs, First Edition. Robert L. Sogge.
© 2012 John Wiley & Sons, Ltd. Published 2012 by John Wiley & Sons, Ltd.

11.3.1 Assumptions

These are a vertical point load, perpendicular to the surface, having a small area of application compared to a semi-infinite body.

A linearly elastic relation is required between σ and ε on loading (rebound not important):

Isotropic strength properties – not direction (orientation) dependent
Homogeneous – one strength property not location dependent
Semi-infinite elastic half-space having weightless mass.

11.4 Westergaard Equation

This is used to solve for the vertical, σ_v, stresses beneath the surface of a layered elastic half-space (semi-infinite weightless mass) due to a vertical point load applied at the surface. The solution includes the effect of Poisson's ratio (Bowles, 1977, p. 153).

11.5 Mindlin Equation

This is used to solve for the vertical and lateral stresses beneath the surface of an elastic half-space (semi-infinite weightless mass) due to a vertical point load applied beneath the surface (Mindlin, 1936, pp. 195–202). The solution is presented in Poulos and Davis (1974, pp. 17–18). This solution reduces to the Boussinesq solution when the vertical distance below the surface is zero.

11.6 Chart Solutions

Complete chart solutions are available in Poulos and Davis (1974), as follows, to determine the stresses and displacements beneath the surface of a semi-infinite mass in an elastic half-space subjected to a vertical load:

Loading	Equation (chart) page of Poulos and Davis (1974)
Point load (Mindlin, 1936)	17–19
Rectangular loaded area (Skopek, 1961)	92
Circular loaded area (Nishida, 1966)	94
Line load (Melan)	27

If the vertical load is applied at the surface, the above solutions reduce to:

Boussinesq (1885)	16
Newmark (1942) and Hall (1940)	77 and 54
Giroud (1970), Fadum (1948), and Harr (1966)	–
Foster and Ahlvin (1954)	43
Lysmer and Duncan (1969)	26

The references noted above are supplied as a label to the type of chart solution and are referenced in the text by Poulos and Davis (1974).

 ## 11.7 Excel Workbook – Lat&VertStress

Excel workbook **Lat&VertStress** with worksheets Point-Line-Strip (11.2), Bous&West (11.3 and 11.4), and Mindlin (11.5) present solutions for the formulations in the sections noted. Features of interest in these worksheets are as follows,

In defining names (Insert > define > name) for the values in a worksheet cell, these names are then common between all the worksheets of a workbook.

Note how in the cells of the Mindlin worksheet the terms in the long equation (formula) for the horizontal stress are separated by spaces for easier checking of conformance to the solution equation.

The use of the $ symbol, as in D39, holds the value of certain terms of an equation when copying formulas to other locations.

 ## 11.8 VBA Program HSpace

Program HSpace was written in FORTRAN by John Lysmer of the University of California at Berkeley (February 1968, 1969) and revised by Robert Pyke (November 1972). The program computes stresses and displacements in an elastic half-space due to a general vertical load applied beneath the surface of a semi-infinite weightless mass. The program also provides solutions for applied horizontal loads, and horizontal and vertical shear loads. **Examples 11.1–11.5** provide application data for program **HSpace**. Its documentation is provided in the Appendix to this chapter.

11.9 Significant Programming Aspects

These are as follows:

- The resulting VBA procedural code required little modification to the original FORTRAN program code written in 1968–1969.
- This program is ideally set up for a data file input and output since, due to the many input variable choices, it is more difficult to fix the location of the input in the worksheet cells. For that reason a sequential file is used to input the data and contain the program solution output.
- The format for the large range in numbers output is conveniently handled using the expression

```
Print #2, Format (S(I), "0.00E+00 ");
```

This output format expression readily accommodates general number output. If the + sign next to the E is not used the output can appear wrongly placed. A couple of blank spaces are used at the end of the format to separate the numbers in the file. The ";" character keeps the printed output on the same line, one after another.

11.10 VBA Program HSpace – Program Documentation

The vertical load may be a:

(a) point load (Mindlin, 1936)
(b) rectangular loaded area (Skopek, 1961)
(c) circular loaded area (Nishida, 1966)
(d) line load (Melan).

For surface loadings of an applied vertical load these solutions reduce to those of:

(a) Boussinesq
(b) Newmark (1942), Hall (1940), Giroud (1970), Fadum (1948), and Harr (1966)
(c) Foster and Ahlvin (1954)
(d) Lysmer and Duncan (1969).

The solution is obtained for area loads by decomposing them into a finite number of point loads and then superimposing them using the formulas in Mindlin's (1936) paper. Point, line, and rectangular loads are special cases of the parallelogram loading. This program adds to the above spreadsheet by providing the additional traction and area loadings besides the vertical point load. As with all programs like this one, the formulas used for stresses and displacements blow up at the point of application of the load.

The Excel workbook **HSpace** contains the VBA macro **Hspace**. The worksheet starts this macro using a command button. The macro reads the input data from a sequential data file and outputs the results to a sequential data file.

The positive coordinate system is oriented with x positive to the right, y positive out of the page and z positive down. The elastic solid is located in the half-space with positive z coordinates. Tensile stresses are considered positive.

Parallelogram Load

The first basic element or loading is a force distributed over a parallelogram in space. This loading includes point loads, line loads, and rectangular loadings as special cases.

The force is defined by its components in the x, y, and z directions.

The parallelogram is defined by the coordinates of one of its corners plus two vectors representing the two sides.

Triangular Load

The second basic element of loading is a linear stress distribution over a triangular area in space.

This loading is defined by the coordinates of the corner of the triangle and the components of the stress vectors at these corners.

Dimensioned Size

The number of parallelogram loads plus three times the number of triangular loads is dimensioned for 100 (N2).

The number of point at which results are to be computed is dimensioned for 100 (N2).

Program Data Input

Program input is read from a sequential data file created using a text editor program such as Notepad. Data should be input according to the following format.

On the worksheet input the drive \location and input filename where the data resides.

Title 1 line
   Title = Problem description title
Elastic Mod, Poisson Ratio 2 values on 1 line
   Soil's Elastic Modulus
   Soil's Poisson's Ratio
No. of (i) Parallelepipedal Loads, (ii) Triangular Loads, (iii) Computation Lines, 3 values on 1
   line

Parallelepipedal Loads = number of Parallelepipedal loads
Triangular Loads = number of Triangular loads
Computation Lines = number of Lines along which stresses and displacements are to be computed.

(a) Parallelepipedal Load, X, Y, and Z, No. of Divisions
   Heading information only – Input data in part (a) only if Parallelogram Loads
   Total Force 3 values on 1 line
   X Y Z components of applied load
   At Point 3 values on 1 line
   X Y Z coordinates of corner of Parallelogram
   1-Side Vector, No. of Divisions 4 values on 1 line
   X Y Z components of endpoint of 1st vector
   (zeros if point load)
   Number of divisions of vector (zero if point load)
   2-Side Vector, No. of Divisions 4 values on 1 line
   X Y Z components of endpoint of 2nd vector
   (zeros if line load)
   Number of divisions of vector (zero if line load)
   The product of the two vector divisions must be < 401

(b) Triangular Load (input data in part (b) only if Triangular Loads)
   X-Stress, Y-Stress, Z-Stress, X-Coord, Y-Coord, Z-Coord, Order of Subdivision
   7 values on 1 line 1 line per corner
   X Y Z – Stress = X Y Z components of stress of corner 1
   X Y Z – Coord = X Y Z coordinates of corner 1
   Order of Subdivision = number of divisions of side of triangle
   (< SQRT 400)
   do same for other 2 corners

(c) Endpoints on lines along which stresses are computed at LN points
   (heading info only)
   X1, Y1, Z1, X2, Y2, Z2, LN 7 values on 1 line
   X1 Y1 Z1 = endpoint coordinates of line or point along which stresses and displacements are to be determined.
   X2 Y2 Z2 = endpoint coordinates of other end of line
   = 0 if point load
   LN = no. of equidistant points along line – include both ends.

Program Output

The program outputs the data to a sequential data file having the same name as the input file with "out" added to its name:

No. of Load Elements
Total No. of Point Loads
No. of Pts at which results requested

Pt X, Y, Z – Coordinates Normal Stresses, Shear Stresses, Displacements
Projections of the Resultant Force on the X Y Z Axis
Projections of the Resultant Force about the X Y Z Axis.

Examples – Input Data Files

The following input data files have been created for examples of program HSpace:

HSIT
HSIP
HSIP1
HSI2T1
HSI2T

 Related Workbooks on DVD

Lat&VertStress with Worksheets: Point-Line-Strip, Bous&West, Mindlin Hspace –
   VBA macro program requiring input data files for examples
Hspace(r3)
Example 11.1 HSIT
Example 11.2 HSI2T
Example 11.3 HSI2T1
Example 11.4 HSIP
Example 11.5 HSIP1

References

AASHTO (2010) *Bridge Design Specifications*, 5th edn, Article 3.11.6.2, American Association of
   State Highway and Transportation Officials, pp. 3–121.
Bowles, J.E. (1977) *Foundation Analysis and Design*, 2nd edn, McGraw-Hill.
Burns, J.Q. and Richard, R.M. (1964) Attenuation of stresses for buried cyclinders. Proceedings
   of the Symposium on Soil–Structure Interaction, University of Arizona, Tucson, pp. 378–392.
Lysmer, J. and Duncan, J.M. (1969) *Stresses and Deflections in Foundations and Pavements*, Depart-
   ment of Civil Engineering, University of California, Berkeley.
Mindlin, R.D. (1936) Force at a point in the interior of a semi-infinite solid. *Journal of Applied
   Physics*, **7** (5), pp. 195–202.
Poulos, H.G. and Davis, E.H. (1974) *Elastic Solutions for Soil and Rock Mechanics*, John Wiley &
   Sons, Inc.
Spangler, M.G. (1936) The distribution of normal pressure on a retaining wall due to a con-
   centrated surface load. First International Conference on Soil Mechanics and Foundation
   Engineering (ICSMFE), vol. 1, pp. 200–207.
Terzaghi, K. (1954) Anchored bulkheads. *Transactions of the American Society of Civil Engineers*,
   119, pp. 1243–1324.

12

Lateral Soil Pressures and Retaining Walls

12.1 Lateral Earth Pressure – Sloped Backfill Acting on Inclined Retaining Wall

The active and passive soil pressures acting on a rigid retaining wall have been developed for the condition of a plane failure surface by Coulomb (1736–1806), Rankine (1820–1872), and Culmann (1821–1872). Here, the dates in parentheses encompass their lives. Coulomb's original equation solution encompassed the parameters of wall inclination from vertical, backfill slope inclination, soil friction angle, and wall friction angle. An example of computing the lateral soil pressure acting against a wall is presented in the workbook **Lateral Soil Pressure**.

Culmann, Engesser, and Poncelet developed graphical solutions to the Coulomb equation using a trial wedge method to develop the active and passive pressures acting on a slope. Their solution included the added parameter of cohesion. When using their equations it should be noted that the passive resistance is significantly overestimated when using plane failure surfaces.

The solution to this problem is given in the workbook **Ka-Gen**. This routine computes the active pressure by graphical (trial and error) means for a soil material that has both ϕ and c (and therefore a wall adhesion) components.

Active case: macro ACA computes $(\alpha_{cr})_a$ associated with the maximum value of P_a by incrementing α from 0 to α_{cr}.

Passive case: macro ACP computes $(\alpha_{cr})_p$ associated with the minimum value of P_p by incrementing α from 0 to α_{cr}. Since Function statements return only one value, the equations are used to recalculate P based on α_{cr}.

Solutions for Soil and Structural Systems using Excel and VBA Programs, First Edition. Robert L. Sogge.
© 2012 John Wiley & Sons, Ltd. Published 2012 by John Wiley & Sons, Ltd.

12.2 Slope Stability

For the special case of a vertical cut with no retaining wall holding it, the solution can be derived for the stability of a slope or embankment with a plane failure surface. The stability is expressed by h_{cr}, the critical height that the slope can stand, as $h_{cr} = (4c/\gamma)\sin\tau\cos\phi/\{1 - \cos(\tau - \phi)\}$ where τ is the angle of inclination from a horizontal of the slope (Taylor, 1948). The inclination of the critical failure plane to the horizontal is $(\tau + \phi)/2$. For a $\phi = 0$ clay soil this expression reduces to $h_{cr} = 4c/\gamma$ $\cot(\tau/2)$. For nearly vertical slopes the plane failure surface assumption is good. For flat or shallow slopes the failure arc is nearly circular and in some cases when $\tau < 30$ passes beneath the toe of the slope.

12.3 Stability of a Vertical Cut

Rankine assumed that the presence of the wall does not influence the state of stress in the soil mass behind the wall. Backfills inclined at an angle i to the horizontal will result in the active pressure being inclined at that angle. Rankine's theory does not incorporate the influence of wall friction angle and assumes a plane failure surface behind a wall.

The stability of a vertical unsupported slope of height h in a cohesive soil can be derived from the Rankine equation

$$\sigma_3 = \sigma_1 \tan^2(45° - \phi/2) - 2c\tan(45° - \phi/2)$$

(Bowles, 1977).

For an element at the bottom of this slope, $\sigma_3 = 0$ and $\sigma_1 = \gamma h$ yield

$$h_{cr} = 2c/\gamma \tan(45° + \phi/2)$$

This solution is appropriate for brittle clays.

For the $\phi = 0$ condition

$$h_{cr} = 2c/\gamma$$

and failure will progress upward from the base of the wall.

By considering the entire failure wedge, writing the Rankine equation for active pressure

$$P_a = \gamma h^2/2 \tan^2(45° - \phi/2) - 2ch\tan(45° - \phi/2)$$

and setting it equal to zero yields

$$h_{cr} = 4c/\gamma \tan(45° + \phi/2)$$

which for a plastic clay with $\phi = 0$ yields

$$h_{cr} = 4c/\gamma$$

This solution works for plastic clays and assumes the entire failure wedge will be engaged.

In a manner similar to looking at the stresses at the bottom of a vertical unsupported wall, a bearing capacity approach yields the following equation for the critical height:

$$h_{cr} = 2c/\gamma\{\tan^3(45° + \phi/2) + \tan(45° + \phi/2)\}$$

which for $\phi = 0$ yields $4c/\gamma$. The graphics for this derivation are shown in the workbook **Ka-Gen**.

12.4 Retaining Wall Movements

Terzaghi (1934) found that there is no difference for a wall which yields by rotation about the bottom and one which yields by translation as far as total pressure acting on the wall. A small movement of $0.0007H$ in either tilting or sliding will develop arching pressure in dense sands. Similarly for dense sands, the minimum pressure will be developed in $0.0027H$ in a tilting wall and $0.005H$ with a sliding wall. All these values are for displacements measured at the center of a wall. Loose sand will require a movement of $0.008H$ at the center of the wall.

Movements of a retaining wall toward and away from a soil that result in passive and active pressures are given in the workbook **Soil Material Property Table**, worksheet labeled Ko, presented in Chapter 10. Typically, horizontal strains of 0.5% will produce active pressures and 5% will produce passive pressures (Lambe and Whitman, 1969).

12.5 Retaining Walls – Factor of Safety

Cast-in-place retaining walls should be designed to maintain the following factors of safety for the associated potential failure modes:

Sliding 1.5
Overturning 2.0
Bearing capacity 3.0
Overall stability 1.5 (within a slipping earth mass).

The workbook **RetWallUSD** computes the stability of a retaining wall backfill with soil as well as checking the strength of the concrete component elements.

Any examples of the use of the workbooks **Lateral Soil Pressure, Ka-Gen**, and **RetWallUSD** are contained in a worksheet of the workbook.

 Related Workbooks on DVD

Lateral Soil Pressure
Ka-Gen
RetWallUSD

References

Bowles, J.E. (1977) *Foundation Analysis and Design*, 2nd edn, McGraw-Hill.
Lambe, T.W. and Whitman, R.V. (1969) *Soil Mechanics*, John Wiley & Sons, Inc.
Taylor, D.W. (1948) *Soil Mechanics*, John Wiley & Sons, Inc.
Terzaghi, K. (1934) Large retaining wall tests, *Engineering News Record*, February 1.

13

Shallow and Deep Foundation Vertical Bearing Capacity

13.1 Shallow Foundations

Shallow foundations are usually continuous strip footings or square or rectangular mat or raft foundations. Either may be very thin in comparison to their lateral extent. In such cases their analysis is given special attention in Chapter 17.

13.2 Vertical Bearing Stress Capacity

The ultimate vertical bearing stress capacity, q_u, of a shallow foundation having a rough base and resting on a clay or sand material based on the shear strength and footing geometry can be computed using the Terzaghi bearing capacity equation (Terzaghi and Peck, 1967)

$$q_u = cN_c + \gamma DN_\theta + \gamma B/2N\gamma$$

The maximum soil pressure under the footing should not exceed the allowable pressure, q_a, that is, the ultimate capacity divided by a factor of safety, usually 2.5–3. The bearing capacity coefficients N_c, N_q, and N_γ are related solely to the friction angle ϕ of the bearing material. In Chapter 10, equations for the bearing capacity factors for a shallow rough foundation were given as

$$(aa) = e^{(\pi \times 3/4 - \phi/2)\tan\phi}$$

$$N_q = (aa)^2/\{2\cos^2(45° + \phi/2)\}$$

Solutions for Soil and Structural Systems using Excel and VBA Programs, First Edition. Robert L. Sogge.
© 2012 John Wiley & Sons, Ltd. Published 2012 by John Wiley & Sons, Ltd.

$$N_\gamma = 1.8(N_q - 1)\tan\phi$$
$$N_c = (N_q - 1)/\tan\phi$$

The bearing capacity factors N_c, N_q, and N_γ for shallow footings having a rough base are given in the Excel workbook **BearingCapFactors**. In the worksheet of this workbook calculated factors are compared to those specified in Table 10.6.3.1.2a-1 of the 2007 AASHTO LRFD Bridge Design Specifications. The Excel workbook **Ftg-BrgCap** uses, internally in its calculations, bearing capacity factor and ϕ relations developed from the equations above.

13.3 Soil Pressure Distribution

In addition to the applied load distribution, the distribution of stress in the soil beneath a footing depends on the stiffness and roughness of the footing, the depth of foundation embedment, and the type of soil present. The soil pressure directly beneath a footing and the displacement pattern of the footing structure patterns, taking into account the footing flexibility, soil type, and depth of footing, have been given by Taylor (1948), specifically Figure 19.18 (Pressures distributions and differential settlements in cohesionless soils) on page 614 and Figure 19.19 (Pressures distributions and differential settlements in highly cohesive soils) on page 615. These patterns are shown conceptually on calculation sheet 5 in the Excel workbook **Ftg-BrgCap**.

13.3.1 Smooth and Rough Footing Bottoms

Both rigid and flexible footings can have either smooth or rough bottoms interfacing with and affecting the soil pressure distribution below them. Rough footings will have a slightly higher bearing capacity than smooth footings.

Elastic solution equations presented in Chapter 9 covered stresses in an elastic half-space. Such equations covered point and distributed loads applied on or below the surface of a semi-infinite mass. The stress patterns are typical of those below a smooth, very flexible footing that would exert little influence on the supporting soil in distributing the load. Poulos and Davis (1974) give elastic solution equations for states where the interface boundaries between the soil and the structure are described as either smooth or rough, as follows:

- For flexible footings – Chapter 3, Distributed Loads on the Surface of a Semi-infinite Mass, pp. 36–91.
- For rigid footings – Chapter 7, Rigid Loaded Areas, pp. 165–182.
- For intermediate footings with interface boundaries between the soil and the structure being described as smooth or rough – Chapter 13, Raft Foundations, pp. 249–268.

13.3.2 Eccentric Loadings

By assuming a linear contact pressure distribution (as was assumed in the Rankine pressure distribution for retaining walls), the soil pressure intensity for eccentric loads is computed using the basic engineering equation for computing stress in a beam–column subject to axial load and bending:

$$\sigma = P/A \pm My/I$$

The eccentric loading is first transformed into an axial load and a moment acting though the center of the footing. Taking a 1 ft wide footing strip of the footing having a width B yields

$$A = B \times 1$$

$$y = B/2$$

$$I = (1)B^3/12$$

Further defining the moment as Pe and setting σ_{min} at an edge to zero yields

$$e = M/P < \pm B/6$$

or in the center third of the footing.

No tension will exist in the soil if the equivalent vertical load is applied within the center one-third of the footing, known as the kernel. Thus the footing size should be selected to accommodate this requirement. If the load is outside the kernel then tension arises over a portion of the slab. At the one-third point a triangular pressure distribution results. The Excel worksheet **Ftg-EccentricLd** computes the stress state in the soil beneath eccentrically loaded footings.

13.3.3 Footing Flexibility

It will be shown later in Chapters 16 and 17 how stress distributions beneath a footing can be computed. Calculations that take into account the flexibility of the footing and determine the footing stiffness factor are shown in this chapter. In general, for now, an equivalent net uniform bearing pressure is used as the footing pressure distribution for point and uniformly distributed loadings. This value is obtained by dividing the load by the footing area. Such a calculation is really only applicable for very rigid foundations and in cases where the actual contact pressure distribution is unknown.

13.4 Settlement-Based Bearing Capacity

The capacity of foundations resting on dense cohesionless soils or firm to hard cohesive soils is rarely controlled by the shear strength or related bearing stress value computed.

Except for very small footings at or near the ground surface, allowable settlement controls their capacity. This control is especially true in the case of large spread footings or mats. Capacity charts giving settlement-based curves are needed for such conditions. Schmertmann, Hartman, and Brown (1978) developed a procedure for including settlement control situations.

A settlement-based bearing capacity relation for shallow foundations resting on cohesionless soils in terms of the SPT blow count N value is: allowable bearing capacity based on a 1 in foundation settlement = $N/10$ (tsf). This relation does not rely on an estimate of ϕ from the N value.

The elastic settlement of shallow foundations like the pressure distributions depend on the flexibility of the footing. In the Excel workbook **Ftg-BrgCap** settlement calculations are provided for the limiting cases of flexible and rigid footings (Bowles, 1977, p. 157). Calculations that account for the footing flexibility are covered later, in Chapter 17 of this text.

 ## 13.5 Excel Workbooks

The following workbooks are attached to the information for this chapter:

BearingCapFactors – bearing capacity factors N_c, N_q, N_γ for shallow foundations including an equation for N_q for deep foundations developed using **CurveFit**
Ftg-BrgCap – bearing capacity of a shallow footing
Ftg-EccentricLd – stress pattern beneath an eccentrically loaded footing
Pier-BrgCap – bearing capacity of a deep foundation
Mat-Pier Found – bearing capacity of a hybrid foundation.

13.6 Deep Foundations

Deep foundation systems generally consist of drilled piers, shafts, or caissons that refer to a concrete structure cast-in-place in a drilled hole and driven piles composed of steel H-type, timber, steel shell driven with a mandrel, or pre-stressed concrete. Such foundations can be designed for situations where adequate depth is required to obtain a larger capacity or where a foundation is required to obtain the depth necessary to resist scour. These piles may be connected by a pile cap consisting of a large mat to make them perform together as a group.

The design criteria should establish the following:

- Minimum pier diameter
- Maximum pier end bearing pressure
- Minimum pier depth below ground surface elevation.

13.7 Capacities Based on Displacement Limits

Lymon Reese and Michael O'Neill have been pioneers in establishing both construction and design methods for drilled shafts (Reese and Wright, 1977a, 1977b; Reese, 1984; Reese and O'Neill, 1974, 1988). In their 1988 publication they developed a bearing capacity formulation for drilled piers such that deflections that occur due to end bearing stresses and skin friction shears are compatible. They modified their theory and equations with a less intuitive approach (O'Neill and Reese, 1999) that has been adopted as a series of equations by the current (2010) AASHTO Bridge Design Specification. The theory and application of their 1988 publication is expressed in this chapter. Another reference giving test results for pier bearing capacities in granular soils in the southwestern United States is Beckwith and Bedenkop (1973).

13.7.1 End Bearing

Ultimate failure in end bearing occurs at a tip deflection of:

Cohesionless soils	$5\%\ B$ (Reese and O'Neill, 1988, p. 259)
Driven piles may have half this 5% value	
Cohesive soils	$3\%\ B$ (Reese and O'Neill, 1988, p. 251)

13.7.2 Skin Resistance

Ultimate failure in skin friction occurs at a tip deflection of:

Cohesionless soils	$0.5–3\%\ B$, approximately $1.66\%\ B$ (Reese and O'Neill, 1988, p. 258)
Cohesive soils	$1\%\ B$ (Reese and O'Neill, 1988, p. 250)

13.7.3 Combined Capacity and Factor of Safety

A factor of safety, FS, on end bearing is chosen to make failure in both the end bearing and skin friction modes occur simultaneously, that is, at the same displacement level:

$$Q_{EB}/Q_{SF} = 5\%/1.66\% = 3$$

where Q_{EB} = ultimate pier capacity in end bearing and Q_{SF} = ultimate pier capacity in skin friction.

The ratio of deflections in end bearing and skin friction for both types of soils is approximately 3.

Using an overall FS of 1.5 yields

$$Q_{allow} = (Q_{EB}/3 + Q_{SF})/1.5$$

The FS on end bearing is now 4.5 and on skin friction it is 1.5. The FS of 3 lowers the end bearing stresses to the point where deflections compatible with those due to skin friction stresses result.

13.8 Capacities Based on Stress Limits

13.8.1 End Bearing

The ultimate end bearing capacities, q_u, are computed using Terzaghi's bearing capacity equation (Terzaghi and Peck, 1967) using the bearing capacity coefficient N_q as developed by Meyerhoff for a "bored pile" (Lambe and Whitman, 1969, p. 501):

$$(q_u)_{EB} = cN_c + gDN_q + gB/2Ng$$

$$Q_{EB} = (q_u)_{EB} \times A_{tip}$$

where

$$A_{tip} = \pi D^2/4$$

For deep foundations the same N_c and N_γ factors are used as for a shallow foundation. N_q changes. Fitting an equation to curves describing the bearing capacity factor N_q in terms of ϕ is useful as the equation acts as a "table" providing a value quickly for a specified ϕ value.

By using the "add trend line" feature of charts within Excel, an equation can be developed. Similarly to Chapter 10, program **CurveFit** can be used to fit an equation to curves giving the bearing capacity factor N_q for deep foundations in terms of ϕ. To implement "add trend line" within Excel right click on the trendline and check display equation and display R^2 value on chart. Program **CurveFit** gives a greater range of 25 types of equations that can be used to fit the data. These equations are presented in Table 10.2.

An equation fitting the curve of ϕ (phi) versus N_q based on Meyerhoff's N_q values for a bored pile foundation is as follows:

From a trendline of data charted on an Excel worksheet:

$N_q = 0.0853{}^{*}EXP(0.2208{}^{*}phi)$ yielding an R^{2}correlation parameter of 0.9892

And a better fit from program **CurveFit** with an R^2 correlation parameter of 0.9998,

$N_q = IF(phi < 25, phi, 8.0126{}^{*}EXP(((phi - 6.7786)\hat{}2)/264.7)).$

The bearing capacity factors N_c, N_q, and N_γ applicable for deep bored pier foundations are given in the Excel workbook **BearingCapFactors**. The workbook **Pier-BrgCap** uses these same bearing capacity factor and ϕ relations internally in its calculations.

13.8.2 Skin Resistance

For the ultimate skin friction capacity, Q_{SF}:

Cohesionless soils	$(\tau_u)_{SF} = \tan\varphi(K_o\gamma H)/2$
where K_o = coefficient of earth pressure at rest	
Cohesive soils	$(\tau_u)_{SF} = \text{adhesion}$
where adhesion = a percentage of cohesion up to 45%	

$$Q_{SF} = (\tau_u)_{SF} \times A_{skin}$$

where

$$A_{skin} = (\pi D H)$$

Thus

$$Q_{SF} = \pi D \tan\phi K_o\gamma H^2/2$$

In an uplift condition, skin friction is the only resistance component present.

13.8.3 Bearing Capacity in Terms of Blow Counts

Bearing capacity has been represented in terms of the SPT blow count N value for sand and silt soils similar to the $N/10$ (tsf) settlement–bearing capacity relationship presented for shallow foundations. For the bearing capacity of deep foundations the following relationship has been used:

$$Q_f(\text{tons}) = 8NA_p + N'A_s/50$$

where
$\quad Q_f$ = failure bearing capacity (kips)
$\quad N$ = average blow count (SPT) within zone 3 diameters below and 8 diameters
$\qquad$ above tip
$\quad N'$ = average blow count (SPT) along shaft
$\quad A_p$ = cross-sectional area of point (ft^2)
$\quad A_s$ = cross-sectional area of shaft (ft^2)
$\quad 8$ = factor for sands (range = 8–12)
$\qquad$ gravels (range = 12–16)
$\qquad$ silts (range = 6)
$\quad 50$ = factor for bored piles (25 = factor to be used for driven piles)

A FS of 3 is recommended with this equation. Due to new definitions of the N value by the FHWA, the N-value approach has not been taken here.

13.8.4 Reduction in Capacity Based on Spacing

The minimum on center pier spacing to achieve these capacities should be 3 diameters (Reese and O'Neill, 1988, p. 233). Thus an "efficiency" factor is 1 for a center-to-center pier spacing greater than 3 diameters.

Based on AASHTO methodology, an axial capacity reduction factor should be applied to shafts with a center-to-center spacing of between 8 and 3 diameters apart. The reduction factor is linear and varies between 1 for 8-diameter spacing and 0.67 for 3-diameter spacing and is given by the following equation:

$$\text{Reduction factor} = 0.0667 \times \text{spacing} + 0.472$$

where the spacing is given in diameters.

The fifth edition of AASHTO has the same linear range of reduction factors for:

- Cohesive soils with pier spacing between 2.5 and 6.
- Cohesionless soils with pier spacing between 2.5 and 4.

13.9 Limitations on Capacities

Meyerhoff made recommendations that the ultimate end bearing be limited to the value calculated when the N_q factor is determined for the following depths (Reese and O'Neill, 1988, p. 227):

ϕ	Limiting depth
28	$3 \times$ diameter
30	$3.5 \times$ diameter
32	$4 \times$ diameter
35	$5.5 \times$ diameter
40	$8 \times$ diameter

Ultimate end bearing and skin friction values for the various soil types are limited to the following values in the referenced publication:

End bearing	Cohesionless soils	$(q_u)_{EB} = 90$ ksf	Reese and O'Neill (1988, p. 257)
	Cohesive soils	$(q_u)_{EB} = 80$ ksf	Reese and O'Neill (1988, p. 248)
Skin friction	Cohesionless soils	$(\tau_u)_{SF} = 4$ ksf	Reese and O'Neill (1988, p. 254)
	Cohesive soils	$(\tau_u)_{SF} = 5.5$ ksf	Reese and O'Neill (1988, p. 243)
Adhesion is limited to 45% of cohesion			

13.10 Load Testing

Higher values than these limits could be established with a load test (Townsend, 1993). High-capacity load tests have been conducted using a sacrificial load cell developed by Osterberg (1984, 1989, 1994), and Osterberg and Hayes (1999) at the bottom of a drilled shaft. Results from pier load testing using this load cell can be used to derive the end bearing and skin friction components of resistance. A test using this cell has been conducted to 34 000 kips on an 8 ft diameter pier, 135.5 ft long, in the clayey sand and gravel soil in Tucson, Arizona (refer to the references on load testing).

Another relatively new method which tests the pile from its top is the Statnamic test (Horvath, 1990). This test has been used to lower limits than with the Osterberg cell. A simple dynamic load test that can readily be conducted in the field after pier construction is one using a pile driving analyzer (PDA) (Goble, Rausche, and Likins, 1993).

13.11 Pier Settlement

Total settlement at allowable load under the pier design loads is approximately equal to 1% (actually 1.66%/1.5) of the pier diameter.

 ## 13.12 Excel Workbook

Pier-BrgCap – bearing capacity of a deep foundation.

This workbook computes the allowable (unfactored) vertical pier load (kips) versus the depth (ft) of pier below the scour level based on input soil parameters for various shaft diameters. The program plots on the worksheet the allowable (unfactored) vertical pier load (kips) versus the depth (ft) of pier below the scour level based on $\gamma_{buoyant} = 53\,pcf$, $\phi = 33°$, $c = 0$, for 2, 2.5, 3, 4, and 5 ft diameter shafts using the charting capabilities of Excel. Total settlement under the pier design loads is approximately equal to 1% of the pier diameter and ranges from $^1/_4$ to $^1/_2$ in.

13.13 Combined Foundations – Shallow and Deep

Systems that combine both shallow and deep foundations are often employed. An example of such a system consisting of a foundation system carrying loads by the independent elements of a mat and by drilled piers is shown. In this system the mat and pier components are rigid structures and it is the soil support for the piers and mat that supply the stiffness to the system and determine the distribution of the load to the structural elements.

The analysis of this system is similar to finding the load carried by the rebar and that by the concrete in a reinforced concrete column subject to an axial load as was done in Chapter 3. The Excel workbook **Mat-Pier Found** computes the loads carried in the slab and pier portion of a combined mat–pier foundation.

Related Workbooks on DVD

BearingCapFactors
Ftg-BrgCap
Ftg-EccentricLd
Mat-Pier Found
Pier-BrgCap

References on Shallow Foundations

AASHTO (2010) *LRFD Bridge Design Specifications*, 5th edn, American Association of State Highway and Transportation Officials.

Bowles, J.E. (1977) *Foundation Analysis and Design*, 2nd edn, McGraw-Hill.

Lambe, T.W. and Whitman, R.V. (1969) *Soil Mechanics*, John Wiley & Sons, Inc.

Poulos, H.G. and Davis, E.H. (1974) *Elastic Solutions for Soil and Rock Mechanics*, John Wiley & Sons, Inc.

Schmertmann, H.H., Hartman, J.P., and Brown, P.R. (1978) Improved strain influence factor diagrams. *Journal of the Geotechnical Engineering Division, ASCE*, **104** (GT8), 1131–1135.

Taylor, D.W. (1948) *Fundamentals of Soil Mechanics*, John Wiley & Sons, Inc., pp. 614–615.

Terzaghi, K. and Peck, R.B. (1967) *Soil Mechanics in Engineering Practice*, 2nd edn, John Wiley & Sons, Inc.

References on Deep Foundations

Lambe, T.W. and Whitman, R.V. (1969) *Soil Mechanics*, John Wiley & Sons, Inc.

O'Neill, M.W. and Reese, L.C. (1999) Drilled Shafts: Construction Procedures and Design Methods. FHWA Report No. IF-99-025, Federal Highway Administration.

Reese, L.C. (1984) Handbook on Design of Piles and Drilled Shafts Under Lateral Load. Report No. FHWA-IP-84-11, July.

Reese, L.C. and O'Neill, M.W. (1988) Drilled Shafts: Construction Procedures and Design Methods. Report No. FHWA-HI-88-042, ADSC-TL-4, Federal Highway Administration.

Reese, L.C. and Wright, S.J. (1977) *Drilled Shaft Manual*, Vol. I, Construction Procedures and Design for Axial Loading, Vol. II, Construction Procedures and Design for Lateral Loading, USDOT, HDV-22, July.

Terzaghi, K. and Peck, R. (1967) *Soil Mechanics in Engineering Practice*, 2nd edn, pp. 107, 341, 347.

References on Load Testing of Deep Foundations

Beckwith, G.H. and Bedenkop, D.V. (1973) An Investigation of the Load Carrying Capacity of Drilled Cast-In-Place Concrete Piles Bearing on Coarse Granular Soils and Cemented Alluvial Fan Deposits. Report No. AHD-RD-10-122, Arizona Highway Department, May.

Goble, G.G., Rausche F., and Likins, F. (1993) *Dynamic Capacity Testing of Drilled Shafts*, ADSC, September/October, pp. 12–15.

Horvath, R. (1990) *Statnamic: An Accurate and Innovative Load Test Method for High Capacity Deep Foundations*, ADSC, March/April, pp. 8–12.

References Associated with the Osterberg Load Cell

Goodwin, J.W. (1993) *Osterberg Cell Sets Record in Kentucky*, ADSC, June/July, pp. 21–22.

Litke, S.S. (1991) *Osterberg Load Cell Finding Applications*, ADSC, June/July, pp. 17–18.

Osterberg, J. (1984) *A New Simplified Method for Load Testing Drilled Shafts*, ADSC, August, pp. 9–11.

Osterberg, J. (1989) *Breakthrough in Load Testing Methodology*, ADSC, November, pp. 13–18.

Osterberg, J.O. (1994) *Capacity of Drilled Shafts in Hard Rock*, ADSC, Part I, February; Part II, March/April, pp. 29–33.

Osterberg, J. and Hayes, J. (1999) *Improving Drilled Shafts from Tip to Top*, ADSC, November, pp. 26–34.

Sumi, O.M., Kishida, H., and Yoshifuka, T. (1996) *Application of the Pile Toe Test to Cast-in-Place and Precast Piles*, ADSC, December/January, pp. 23–28.

Townsend, F.C. (1993) *Comparison of Deep Foundation Load Test Methods STGEC'93*, ADSC, November, pp. 20–26.

14

Slope Stability

In the conventional or ordinary method of slices (sometimes denoted as the ordinary methods of slices) developed by Fellenius, the resultant of the forces acting on the sides of each slice is assumed to act parallel to the base for each slice and the sum of these forces is zero when all slices are considered. Such an assumption is true for an infinite slope. Bishop's method of slices assumes that the forces acting on the sides of any slice have zero resultant in the vertical direction (Bishop, 1955).

Slopes having inclination angles greater than $53°$ are steep and generally fail through the toe of the slop. Shallower slopes fail in a deep-seated manner.

These programs employ input from and write their output data to sequential data files. The input data drive\location and name are read from worksheet cells so that the macro does not have to be edited to input this data. The output filename is given the input name with OUT added to the input filename. This change is accomplished automatically by stripping off the four-character extension.xxx and adding OUT to the name input. An interior "." does not create any problem with the input.

A finite element analysis can be performed on a failure embankment. The available and developed shear forces along an assumed failure arc through the slope are summed for each of the elements the arc passes through. Their ratio then determines the factor of safety of the slope. The results of a finite element analysis of a sloped embankment show that the stress in the vertical direction is close to a function of depth in the slice $k_y = \sigma_v/(\gamma z) = 1$ to 1.2 where z is the vertical distance from the slope surface. A much more extensive examination of the subject of slope stability is offered in the text by Duncan and White (2005).

14.1 Workbook Program Slope – Slope Stability by Bishop's Modified Method of Slices

Program Slope was created by John Cross of STS Consultants in the BASIC language and published in the October 1982 issue of *Civil Engineering Magazine*. This program

Solutions for Soil and Structural Systems using Excel and VBA Programs, First Edition. Robert L. Sogge.
© 2012 John Wiley & Sons, Ltd. Published 2012 by John Wiley & Sons, Ltd.

determines the factor of safety of circular slip failure surfaces through slopes for specified circle centers using Bishop's simplified method of slices. The factor of safety can be defined as the ratio of the resisting force moment to the driving force moment about the failure arc center. A total or effective stress approach, with or without a pseudo-static seismic analysis, can be performed.

The minimum factor of safety, FS, should be obtained by varying the location of circle centers and their radii. The FS computed by this program for **Example 14.1** SIEP of 1.91 does not agree with the FS presented in the article (1.96) since this program applies the pseudo-static earthquake force at the base of the slice rather than at its midpoint.

Either a total or an effective stress analysis can be performed by the program. A rapid drawdown analysis uses total weights and total strength parameters for the stress condition existing before drawdown. The long-term steady-state analysis with no seepage uses either of the following approaches:

(a) total weights above and below a specified water table;
(b) effective weights for soils below where the water table would exist without specifying a water table.

Effective strength parameters should be used in both approaches. For submergence using approach (a), the slices are taken through the water. The weight of the water above and below the exit of the failure surface from the slope is included in the slice weight and driving force.

Example 14.1 SIEP provides data for an application of this program. Sequential data files are used for inputting data to the program and for receiving its output. The program documentation is a separate file on the attached program disc.

 ## 14.2 Workbook Program STABR – Slope Stability by Bishop's Modified Method of Slices

This file provides documentation for program STABR written by Guy Lefebvre of the University of California at Berkeley in 1971. The program calculates the factors of safety for specified circles, or searches for the circular slip surface having the minimum factor of safety, using either the ordinary method of slices or Bishop's modified method. The program can be used for total or effective stress analyses, or a combination of both, and with or without seismic forces. The program is capable of handling irregular slope profiles, tension cracks, soil layers with different properties and non-uniform thickness, complicated pore pressure patterns, and irregular variations of undrained strength with depth.

Example 14.2 SRIEP provides data for an application of this program. The program documentation is a separate file on the attached program disc.

14.3 Workbook Program Slope8R – Slope Stability by Spencer's Procedure for Non-circular Slip Surfaces

Slope8R was originally written in FORTRAN by Stephen G. Wright in 1969. It calculates the factor of safety for specified non-circular slip surfaces by the procedure developed by Spencer (1967) and extended by Wright to non-circular surfaces. This is a very stable numerical procedure. By assuming that all side forces are parallel, this procedure satisfies all equilibrium conditions for each slice. The two unknown parameters, F (the factor of safety) and THETA (the side force inclination) are varied simultaneously by iteration until a convergent solution is found with net force and moment imbalance less than specified values.

Although the solution presented by Spencer was only directly applicable to a circular shear surface, his procedure may be readily extended to slip surfaces of a general shape. Spencer's procedure of analysis satisfies all conditions of equilibrium and may be used to obtain a unique solution. The assumption of parallel inter-slice forces may not always lead to the most reasonable solution as judged from the calculated line of thrust. It is interesting to note that if the solution for the side force inclination is zero (horizontal side forces only) then Spencer's and the modified Bishop's procedures are identical and they will satisfy complete equilibrium.

Example 14.3 S8RIEP provides data for an application of this program. The program documentation is a separate file on the attached program disc.

 ### Related Workbooks on DVD

Slope – VBA macro program requiring input data files for examples
Example 14.1 SIEP
Slope-Documentation
STABR
Example 14.2 SRIEP
STABR-Documentation
Slope8R
Example 14.3 S8RIEP
Slope8R-Documentation

References

Bishop, A.W. (1955) The use of the slip circle in the stability analysis of slopes. *Geotechnique*, **5** (1), 7–17.

Cross, J. (1982) Slope stability program. *Civil Engineering – ASCE*, **52**, 71–74.

Duncan, J.M. and Wright, S.G. (2005) *Soil Strength and Slope Stability*, John Wiley & Sons, Inc.

Lefebvre, G. (1971) PhD Dissertation, University of California, Berkeley.

Spencer, E. (1967) A method of analysis of the stability of embankments assuming parallel inter-slice forces. *Geotechnique*, **17**, 11–26.

Wright, S.G. (1969) A study of slope stability and the undrained shear strength of clay shales. PhD Dissertation, University of California, Berkeley.

15

Seepage Flow through Porous Media

15.1 Program Flownet for Analysis of Seepage Flow through Porous Media

The finite element program **Flownet**, written in VBA, solves the modified Laplacian-type partial differential equation governing seepage through a general anisotropic and non-homogeneous saturated porous medium at a finite number of nodal points throughout a domain by formulating and minimizing the associated energy functional expressed in terms of either a potential flow function, Ψ, or a potential head function, Φ, at the nodes connecting the triangular elements into which the flow domain is idealized (Desai, 1972, 1977; Zienkiewicz, 1966, 1971).

Input data for this program consists of nodal point coordinates, element connectivities, material properties described by the principal permeabilities and their orientation with respect to the global axes, and the geometric or forced boundary conditions consisting of the flow or head values along the boundaries depending on whether the potential flow or head function is being determined.

Output consists of the potential flow and head functions at the nodes from which flow and equipotential lines are plotted by the program. These programs employ input from and write their output data to sequential data files. The input data drive\location and name are read from worksheet cells in a similar manner as was done with the programs in Chapter 14 on slope stability. The old plots can best be removed by using Find and Select to Select Objects and then delete.

The program has the capability of handling media having anisotropic flow characteristics by specifying different principal permeability values for each material type along with their orientations to the global coordinate system. Non-homogeneity is handled by specifying different material permeability properties for various zones of elements within the flow domain.

Solutions for Soil and Structural Systems using Excel and VBA Programs, First Edition. Robert L. Sogge.
© 2012 John Wiley & Sons, Ltd. Published 2012 by John Wiley & Sons, Ltd.

15.2 Program Input – from Data file

(1) Title 1 line
   Title = Problem description title
(2) No. of: (a) Nodes, (b) Elems, (c) Matls, (d) Flow BC, (e) Head B C
   5 values on 1 line
   Nodes = number of nodes
   Elems = number of triangular elements in media
   Matls = number of different material types in media
   Flow BC = number of flow boundary conditions
   Head BC = number of head boundary conditions
   (a) Node, X-Coordinate, Y-Coordinate
       3 values/line. As many lines as no. of nodes
       Node = number designating the node
       X-Coordinate = X-coordinate of node
       Y-Coordinate = Y-coordinate of node
       Positive coordinate system is as shown in Figure 15.1.
       The number of equations equals the number of nodes. The maximum node separation
       (Node Sep) on any one element and the bandwidth of the simultaneous equations
       are calculated. Numbering of the nodes should be performed so as to make the node
       separation for nodes on any one element as small as possible, or alternatively, a node
       separation minimizer program should be used.

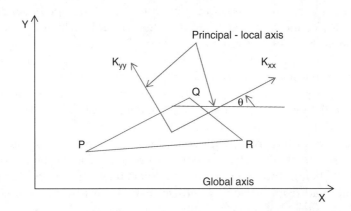

Figure 15.1   Positive coordinate system.

   (b) Element, P-Node, Q-Node, R-Node, Matl Type
       5 values/line. As many lines as no. of elements
       Element = number designating the element
       P-Node = number of P node of triangular element corner
       Q-Node = number of Q node of triangular element corner

R-Node = number of R node of triangular element corner
Matl Type = Material Type of the element
The node designation is sequential in a clockwise direction on these triangular elements.
(c) Matl Type, Prin Kxx, Prin Kyy, Prin Dir-deg
4 values/line. As many lines as no. of materials
Matl Type = number designating the material
Prin Kxx = Principal permeability in x direction
Prin Kyy = Principal permeability in y direction
The ratio of the magnitudes of the principal permeabilities, not their absolute values, is important.
Prin Dir-deg = angle measured (Theta), positive counterclockwise, of the PX principal permeability direction from the global (X) axis direction of the system as shown in the following:.

Permeabilities of soil types	cm/s
Sand	$10^{-0} - 10^{-3}$
Silt	$10^{-3} - 10^{-6}$
Clay	$10^{-6} - 10^{-9}$

(d) NODE, FLOW BOUNDARY CONDITION
2 values/line. As many lines as no. of flow BC
Node = node number
Flow boundary condition = value of flow function at the specified boundary node
(e) Node, head boundary condition
2 values/line. As many lines as no. of head BC
Node = node number
Head boundary condition = value of head function at the specified boundary node.

15.3 Program Output – to Data File

Units Output = Units Input
No of Eqns, Bandwidth, Node Separation
Node, Flow Boundary Condition
Node, Potential Flow Function
Node, Head Boundary Condition
Node, Potential Head Function.

The program input–output is saved in a sequential data file named by the user.
The solution is dimensionless so any consistent value of input units may be used.

15.4 Input Data Description

The number of elements which can be analyzed is unlimited as long as the other restrictions are satisfied since their information is stored in a data file and only called when needed. This procedure along with the ReDimensioning of variable sizes allows more variable storage space within the available memory for the simultaneous equations.

A coordinate transformation from the global to the local system corresponding to the principal permeability directions is performed by the program. Since scalar quantities are added at each node they are equally valid for any orientation of the local axis.

The boundary condition extremes should be, for convenience, between 0 and 100 and at most between 0 and 10 000. The head boundary conditions are input just after the flow boundary conditions. The output functions are calculated immediately after input of the corresponding boundary conditions.

Natural or free boundary conditions are automatically satisfied by the formulation. Therefore the input of the gradient of the boundary conditions is not required.

15.5 Output Data Description

The output consists of the flow function values and head function values at each of the nodes. The flow function values are output before the head boundary conditions are read in and the head value calculated. These values have the same range as those input in the boundary conditions. From these values at each node, contours of equipotential (head) and flow lines are plotted. These equi-function lines are perpendicular to each other for the case of flow through a homogeneous and isotropic medium. Since the interpolation of the contours is linear it is possible that, because of the choice of element node points, a corner of the mesh may be cut across as the outside equipotential or flow line.

The location of the top flow line on a dam can be evaluated by assuming a location within the mesh. The head potentials along the chosen top flow line are assumed unknown. If the potentials evaluated along the dam face do not provide a phreatic surface which has no flow perpendicular to it and no pressure on it, then a second approximation to the top flow line should be chosen until, by trial and error, convergence is achieved. Since this process is tedious and involves the creation of new meshes at each stage, it is easiest to construct a top flow line by a Casagrande-type construction and use it as a first approximation (Finn, 1967; Kealy and Busch, 1971; Taylor and Brown, 1967).

 ## 15.6 Example

Program instructions can most readily be employed by executing the input and output data file for **Example 15.1 FNISP2** on the DVD supplied with this text. During execution the program reads the input data file and writes it exactly as it was input. A scratch input–output file (#3), is used to store the computed potential flow and head functions at each of the nodes for input later to the plot subroutine. This data file is automatically created on the same drive/location that the output data is sent to. It is named, in this case, **Example 15.1 FNISP2Elem**, based on the input data file. Since the solution is dimensionless the input data units selected are arbitrary. The solution to this problem as presented on page 112 of Cedergren (1967) is presented in **Example 15.1**. The input mesh and boundary conditions are shown in Figure 15.2a and b. The numeric flow and head potential output and the plot of Flownet is presented in the workbook example. Symmetry is used in obtaining the solution. For areas of the mesh where the flow or equipotential lines are closely spaced or are rapidly changing, a fine mesh gradation should be used. A coarse mesh can be used at other locations. In the creation of a proper model the determination of an adequate number of elements must fulfill the requirement that any increase in the number of elements or extent of the mesh results in

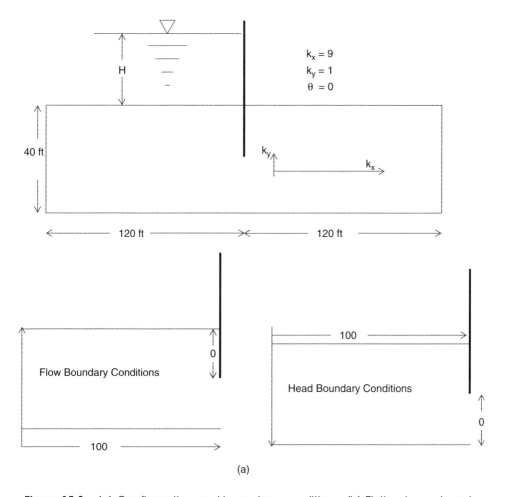

(a)

Figure 15.2 (a) Configuration and boundary conditions. (b) Finite element mesh.

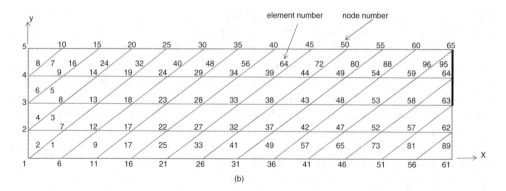

Figure 15.2 *(continued)*

changes in the maximum output parameter of less than 10%. If, when adding elements, the output parameters do not change significantly then the model is adequate. This simple example did not require or employ a variation in mesh size.

The flow and head potential contour lines are constructed. In this example the "squares" are distorted due to the anisotropic permeability properties. For the given example $q/(H \times \gamma(k_h \times k_v)) = 0.5$, or the ratio of the number of flow lines to the number of equipotential lines for the entire flow regime is 0.5.

15.7 Significant Aspects of Excel Workbook and VBA Macro Program Construction

The workbook **Flownet** clarifies the nomenclature in that it has a module that contains macros of Sub or Function Procedures written in VBA. Both the Sub and Function Procedures are written using the same VBA statement language.

The workbook **BatchtoSpreadsheet** converts data in a sequential text file form to worksheet cell data for input to the VBA program **Flownet**. To get by the problem of extracting comma-delimited data from a line of data ended by a carriage return/line feed, it reads the data line at a time using the Line Input # statement and then outputs that single line of data to a file. Then data out of each file with one line of data is read sequentially. An EOF function is used to insure that the program is not attempting to read data past the end of either the large data file or an individual line in a file. The variable A is specified as a Variant so the data can be of any type. Where the file data consists of words it can have spaces between letters and is read in as a string. Where the file data is numbers either commas or spaces can be used to separate them.

This workbook and the associated VBA Plot Sub Procedure show two ways to graphically display the flow net. In one way, the VBA Plot Sub Procedure outputs the plotted flow and equipotential lines to an area on the Excel worksheet labeled "FN" by the following statements associating the worksheet to the VBA macro:

```
Dim myDocument
Sheets("FN").Select
Set myDocument = Worksheets(1).
```

Another approach to graphically display the results is to output the potential head and flow function values to the worksheet. In the worksheet these values are charted using the Excel Chart utility to plot the flow net. Every time a new set of data is entered in the cells associated with the charting, a new chart is created. Therefore there is no need to clear the chart workspace before plotting another.

 Related Workbooks on DVD

Flownet – VBA macro program requiring input data files for examples
Example 15.1 FNISP2
BatchToSpreadsheet

References

Cedergren, H.R. (1967) *Seepage, Drainage and Flow Nets*, John Wiley & Sons, Inc.

Desai, C.S. (1977) Flow through porous media, in *Numerical Methods in Geotechnical Engineering* (eds. C.S. Desai and J.T. Christian), McGraw-Hill, pp. 458–505.

Desai, C. and Abel, J. (1972) *Introduction to the Finite Element Method*, Chapter 12, Van Nostrand Reinhold.

Finn, W. (1967) Darcy flow solutions with a free surface. *Journal of the Soil Mechanics and Foundations Division, ASCE.*, **93** (SM6), 41–48.

Kealy, C. and Busch, R. (1971) Determining Seepage Characteristics of Mill-Tailings Dams by the Finite Element Method. Bureau of Mines, Report of Investigations 7477, U.S. Department of the Interior.

Taylor, R. and Brown, C. (1967) Darcy flow solutions with a free surface. *Journal of the Hydraulics Division, ASCE.*, **93** (HY2), 25–33.

Zienkiewicz, O. (1971) *The Finite Element Method in Engineering Science*, Chapter 15, McGraw-Hill.

Zienkiewicz, O., Mayer, P., and Cheung, Y. (1966) Solution of anisotropic seepage by finite elements. *Journal of the Engineering Mechanics Division, ASCE.*, **92** (EM1), 111–120.

Part Four
Soil–Structure Interaction

"It would be possible to describe everything scientifically, but it would make no sense; it would be without meaning, as if you described a Beethoven symphony as a variation of wave pressure."

– Albert Einstein

16

Beam-on-Elastic Foundation

16.1 Theory – Classical Differential Equation Solution

In considering a beam resting on an elastic foundation, the soil can be modeled as an elastic solid using stress–strain relations as was seen in Chapter 11. Alternatively, the differential equation used by Hetenyi (1946) in his theory and the soil model he employed could be used. In the Hetenyi approach the soil was represented by a Winkler foundation where the springs are independent of each other and provide a stiffness represented by the coefficient of subgrade reaction, k, in the pressure–deflection relation $p = ky$.

Since any finite element solution is merely a different approach to getting an elastic solution other than solving difference or differential equations, it is beneficial to examine the governing differential equation to see the parameters governing beam-on-elastic foundation behavior (Seely and Smith, 1952; Timoshenko and Woinowsky-Kreiger, 1959).

The equation of the elastic curve is

$$EI\,d^2y/dx^2 = -M$$

By double differentiation of the beam theory equation the following differential equation results:

$$EI\,d^4y/dx^4 = q - p = q - ky$$

where
  $x =$ distance along beam or plate
  $y =$ deflection of beam or plate
  $E =$ elastic modulus
  $B =$ width of beam or plate
  $I = Bt^3/12$ for beam section moment of inertia – if plate $t^3/\{12(1 - v^2)\}$ can be used for moment of inertia of a unit-wide elemental strip

Solutions for Soil and Structural Systems using Excel and VBA Programs, First Edition. Robert L. Sogge.
© 2012 John Wiley & Sons, Ltd. Published 2012 by John Wiley & Sons, Ltd.

M = bending moment in beam
q = load intensity acting on beam or plate
p = soil pressure reaction = ky
k = coefficient of subgrade reaction per unit area.

For a strip of unit width, $B = 1$, and B drops out of the beam equations.
The elastic equation assumes:

- linear thin-plate theory for small deflections, δ, in which $\delta <$ thickness/3;
- least horizontal dimension/thickness > 5 (least horizontal dimension is the distance from the loaded point to the plate edge – this theory does not work well for plates with corner loads).

Regardless of the loading on the beam or plate, the Hetenyi (1946) solution for the beam-on-elastic foundation differential equation contains a characteristic or non-dimensional constant that is a ratio of the soil stiffness to the structural stiffness:

$$S = 4\{(BkL^4)/(4EI)\}^{0.5}$$

The soil stiffness is proportional to $(Bk)^{1/4}$, where k is the coefficient of subgrade reaction of the foundation strip of B. The structural stiffness is proportional to $(4EI/L^4)^{1/4}$ where E is the elastic modulus of the beam, I the moment of inertia of the strip of width B of the beam, and L is the beam's length

16.2　Beam–Bar Finite Element Model

The model used to analyze the moments in the slabs due to the imposed loads consists of the Hetenyi (1946) representation using the Winkler soil model. Beam members or elements represent the slab. The soil is represented as a series of springs or bar elements. Their strength is independent of adjacent spring elements and is prescribed by the coefficient of vertical subgrade reaction, k. A beam of length, L, having the strength properties, EI, of the concrete slab, represents the slab. The model shown in Figure 16.1 represents the beam–bar finite element model (Bowles, 1974). The resistance (stiffness) of the soil continuum in the foundation (soil) is proportional to the deflection of the beam at that point and is independent of the pressure or deflection occurring at other parts of the foundation.

The force–deformation relations, $F = [S]\delta$, for bar members representing the uniaxial soil resistance, are developed similar to the general theory presented in Chapter 3 previously using relations based on stress and strain.

For equilibrium of a bar element, the internal element forces, F, are related to the bar (soil) stresses, p, by

$$F = Ap \tag{16.1}$$

where A = the contributory contact area of the soil.

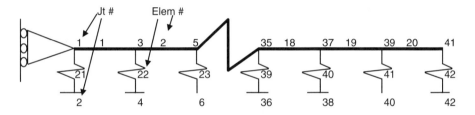

Figure 16.1 Beam–bar finite element model.

Stress–strain, p–ε, relations for the bar are

$$p = E\varepsilon \tag{16.2}$$

where E = an elastic modulus.

Geometric compatibility or strain–deformation is given by

$$\varepsilon = \delta/l \tag{16.3}$$

where l = the bar element length, not to be confused with L, the beam element length. Combining Equations 16.1–16.3 yields the familiar

$$F = (AE/l)\delta \tag{16.4}$$

which can be rewritten as

$$(F/A) = E(\delta/l) \tag{16.5}$$

For soil, the relation between pressure, p, and displacement, δ, on the system level is given by

$$p = k\delta \tag{16.6}$$

where k = the coefficient of subgrade reaction.

If the length of the bar is taken as unity in Equation 16.5, then a comparison of Equations 16.5 and 16.6 shows that E can be replaced by k. The stiffness relation for a bar becomes

$$F = (Ak)\delta \tag{16.7}$$

The constitutive equations for this model consist of coefficients formed by the addition of the beam element stiffnesses and bar element stiffnesses. For these independent spring resistances, the bar element stiffnesses form a diagonal stiffness array. In developing the stiffness array relating loads to deflections in the finite element method, $P_i = S_i y_i$, where S is a diagonal matrix. This independence in the relation of load and deflection is a result of the Winkler assumption of independent springs. Numerically the soil stiffness value can just be added into the diagonal terms of the structure's K matrix. This approach makes for a very small bandwidth. If instead of a Winkler foundation an elastic solid were used, the entire stiffness matrix would become populated.

As noted in the Chapter 7, springs, or bar elements, can be developed in any of three ways. The most efficient way is to use the bar element that consists of a beam element pinned on both ends (member type $= 3$) which is in program **PFrame** to directly represent the spring. To implement a finite element solution of a beam-on-elastic foundation problem the beam should be idealized as a strip of some width, B, and whose length, L, is divided into approximately 15–20 members. With a frame solution, the number of beam elements required is directly evident from looking at the frame to be modeled. In the creation of a model involving soil–structure interaction (SSI) situations, the determination of an adequate number of elements must fulfill a requirement that any increase in the number of elements results in changes in a maximum moment or stress parameter of less than 10%. If, when adding elements, the maximum moment does not change significantly then the model is adequate.

The soil will be represented by bar members having an elastic modulus equal to the coefficient of subgrade reaction, k. The contributory area, A, of the bar element equals the contact area of the soil supporting the beam. This area equals the width of the strip, B, times a length equal to the sum of half the distance on each side of a joint, usually $L/$(number of elements). The length of the bar, L, has been set equal to unity. Therefore E and k are equivalent, though not in units, in the formulation.

16.3 Soil Strength – Coefficient of Vertical Subgrade Reaction

The strength properties of a soil in vertical bearing are represented by the coefficient of vertical subgrade reaction, sometimes called the modulus of vertical subgrade reaction. This strength value relates the soil pressure to deformation. This coefficient is:

- nonlinear;
- confining pressure dependent; and
- depends on the width of the structure.

Values of the coefficient of subgrade reaction, k, are not determined directly in a laboratory test. Instead, they are derived from the measurement of the applied vertical load and deflection of an instrumented plate conducted in the field. For cohesionless (sand) soils the value of k is derived from 1 ft square plate load tests (Terzaghi, 1955). The soil properties for soil stresses that are less than $1/3$ to $1/2$ of ultimate conditions can be sufficiently approximated by linear initial tangent or secant relations between pressure and displacement. Systems loading the soil into its yield state would require an incremental or iterative analysis using tangent values of k that are stress-level dependent.

Values of k_{11} for sand materials are presented in Table 10.1. For sands whose strength is confining pressure dependent, k values decrease for increasing structure widths due to the increased size of the pressure bulb. The relation between k for a beam of general width B and k_1 for a beam 1 ft wide, or k_{11} of a 1ft × 1ft square plate, is

$$k_1 = k_{11}\{(B+1)/2B\}^2$$

In the limit as B becomes very large, this expression becomes $1/4$.

Values of k_{11} for clay materials, whose strength is somewhat independent of depth, are also presented in Table 10.1. The data in this table can be approximated by $k_1 = 50q_u$. For clay materials the coefficient of vertical subgrade reaction, k, for a beam of width B, is $1/B$ times that coefficient for a beam having a width of 1 ft, k_1, or $k = k_{11}/B$. Incorporating beam length, L, yields

$$k = k_{11}(L/B + 0.5)/(1.5L/B)$$

which for long beams reduces to $k = k_{11}/(1.5B)$.

Another method of determining k values is by relating them to soil strength properties such as E that can be determined in a laboratory. This relation is evident in the presentation of k values given in Table 10.1. As shown in this table, k (kcf) is approximately equal to E (ksf).

16.4 Structural Stiffness

As discussed in Chapter 8, the reduced stiffness of a concrete section that occurs after cracking begins can be derived using:

- The Ghosh *et al.* (1996) equations in Table 8.2
- ACI 318-89 Building Code Section 10.11.5.2, equations 10.10 and 10.11
- Program PMEIX developed by Reese (1984).

16.5 Soil–Structure Interaction

The moments in the structure and pressures in the soil are dependent on a non-dimensional ratio, S, of the soil stiffness, Bk, to the structural stiffness, $4EI/L^4$. This ratio is the characteristic of the governing differential equation set up by Hetenyi (1946):

$$S = \sqrt[4]{BkL^4/(4EI)}$$

This characteristic for a 1 ft wide strip, where the B term drops out of the soil stiffness and the I terms, determines the following ranges for how the foundation slab behaves in comparison to the supporting soil:

$$\sqrt[4]{kL^4/(4EI)} < \pi/4 \text{ rigid slab or very weak (soft) soil}$$

for which moments, displacements, and soil pressure distributions can be computed based on statics;

$$\pi/4 < \sqrt[4]{kL^4/(4EI)} < \pi \text{ intermediate soil and structural stiffness}$$

for which the interaction present between the soil and structure can best be determined using a beam–bar finite element model;

$$\sqrt[4]{kL^4/(4EI)} > \pi \text{ flexible slab or very stiff soil}$$

for which soil pressures equal the loading pressures applied.

16.6 Unbalanced Fixed-End Moment from Triangular Load Distribution

The application of a distributed triangular load increasing in intensity results in an unbalanced fixed-end moment (FEM) on any joint equal to $wl^2/15$. The incremental distance between each joint is l and w is the distributed load intensity at the joint.

An evaluation of the importance of using an accurate distributed load versus applying the distributed load as a series of concentrated point loads at the joints can be made by a comparison of the moments and deflections for a beam-on-elastic foundation system loaded by the two different loadings. Such a comparison is made in **Examples 16.1a** and **b**. As can be seen from these results, the deflections are identical and the developed moment in the point load approximation of the distributed load is exceedingly small (0.34 k-ft). Thus little is gained by applying the unbalanced FEM to the structure in terms of achieving a more accurate result.

16.7 Pressure Distribution

The pressure distribution, which exists beneath a footing as well as the maximum moment in the foundation, depends on the S ratio. A typical plot for bending moment in a slab as a function of the soil–structure stiffness ratio is presented in Figure 16.2.

For the case of a flexible mat on a cohesionless soil and mats on soft clay cohesive soils, the Winkler assumption of independent springs is good. For very flexible footings (stiff soils) a uniform soil pressure distribution will result.

For cases like a rigid mat located on the soil surface where sloughing at the edges occurs, the actual stress distribution is known (see Section 13.3). For a rigid footing (weak soil) a distribution typical of elastic homogeneous isotropic soil media results, with increased intensities at the edges decreasing toward the center. Verification of the beam-on-elastic foundation model using program **PFrame** has been accomplished by performing elastic analyses on problems having known classical solutions such as footings on elastic media.

Example 16.1   (a) Triangular point loads on foundation. (b) Triangular distributed loads on foundation.

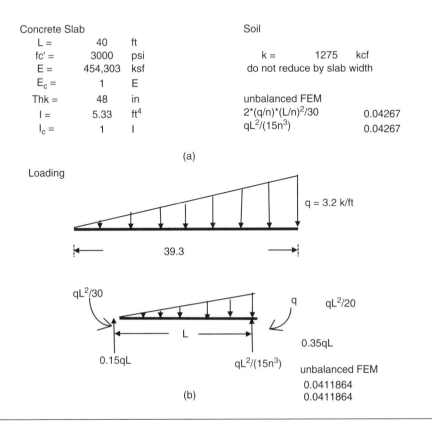

Concrete Slab

L =	40	ft
fc' =	3000	psi
E =	454,303	ksf
E_c =	1	E
Thk =	48	in
I =	5.33	ft⁴
I_c =	1	I

Soil

| k = | 1275 | kcf |

do not reduce by slab width

unbalanced FEM

| $2^*(q/n)^*(L/n)^2/30$ | 0.04267 |
| $qL^2/(15n^3)$ | 0.04267 |

(a)

Loading

$q = 3.2$ k/ft

39.3

$qL^2/30$

$qL^2/20$

q

L

0.35qL

0.15qL

$qL^2/(15n^3)$

unbalanced FEM

0.0411864

0.0411864

(b)

16.8 Solution Exclusively in Excel Worksheet without VBA

Setting up the equations that need to be solved for a beam-on-elastic foundation model can be performed exclusively within the worksheet without using the VBA program **PFrame**. For a simple problem having a generalized configuration of beams and springs, the spring stiffness can be added to the diagonal of the S matrix. The function MINVERS inverts the S matrix and develops deformations. The member forces are computed using the math functions MMULT available within Excel.

When using the math function MINVERSE from the Insert > Function > Math and Trig menu bar in Excel to invert a matrix, do not enter braces { }. Matrix algebra operations such as Transpose, Multiply, and Matrix Inversion cover arrays defined by a range of rows and columns. First highlight the destination matrix and then enter the =MMULT or =MINVERSE functions with the appropriate arrays. This function is

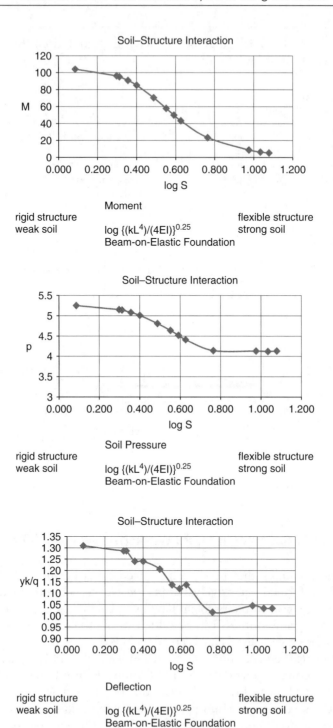

Figure 16.2 SSI curve for beam-on-elastic foundation.

entered by pressing Ctrl+Shift+Enter simultaneously. The braces will automatically be entered during the ShiftCtrlEnter step. They indicate an array formula.

Example 16.2b (Crusher Mat Point Loads) illustrates these procedures.

Example 16.2 **(a) Crusher mat point loads – PFrame solution. (b) Crusher mat point loads – Excel solution.**

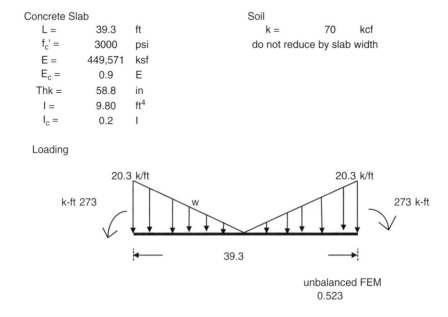

Concrete Slab			Soil		
L =	39.3	ft	k =	70	kcf
$f_c' =$	3000	psi	do not reduce by slab width		
E =	449,571	ksf			
$E_c =$	0.9	E			
Thk =	58.8	in			
I =	9.80	ft^4			
$I_c =$	0.2	I			

Loading

20.3 k/ft   20.3 k/ft

k-ft 273   w   273 k-ft

39.3

unbalanced FEM
0.523

16.9 Examples

16.9.1 Example 16.1a Triangular Point Loads on Foundation; Example 16.1b Triangular Distributed Loads on Foundation

These two examples show two ways of applying a distributed triangular load distribution to the model. One way applies a series of joint loads to the joints of the model and another applies the load as a distributed member load. The results are similar. Another observation from this comparison is that in applying solely joint loads the unbalanced FEM acting on the joint is ignored. For this model that amounts to ignoring a moment application at each joint of $qL^2/(15n^3)$ where q is the maximum distributed load intensity, L is the beam length, and n is the number of members comprising the beam. For this load and model configuration the ignored FEM is 0.043 k-ft, a very small quantity compared to the 3.2 k/ft load. For this reason distributed loads will often be replaced by point loads in the succeeding chapters where distributed soil pressures are considered.

16.9.2 Example 16.2a Crusher Mat Point Loads – PFrame Solution; Example 16.2b Crusher Mat Point Loads – Excel Solution

These two examples show two different ways of solving a problem. The (a) example shows the usual way using program **PFrame**. The (b) example shows the use of the INVERSE and MMULT functions available in Excel. It sets up the constitutive equations and solves them direction using the Excel functions. The same result is achieved even though the (a) model uses 20 elements and the (b) model uses 16 elements. The unbalanced FEM on this model that arises due to applying point loads rather than a distributed load is 0.52 k-ft, a small number compared to the 20.3 k/ft loading intensity.

16.9.3 Example 16.3 SSI Data Generation Model

This example employs the model used for the creation of Figure 16.2, the SSI plot of moment and soil pressure versus the log of the flexibility number S for concrete slabs of various thicknesses. The relation for the moment and soil pressure developed between stiff structures on weak soils and flexible structures on strong soil is quantified. Plots of the beam displacement, moment, shear, and soil pressure output are shown in the bottom of the worksheet for this example.

 Example 16.3   SSI data generation model.

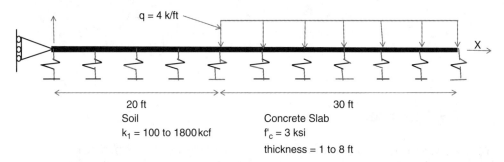

 Related Workbooks on DVD

Example 16.1a Triangular PtLds
Example 16.1b Triangular DistLds
Example 16.2a Crusher Mat – PtLds – PFrame Solution
Example 16.2b Crusher Mat PtLd – Excel Solution
Example 16.3 SSI Beam-on-elastic foundation

References

Bowles, J.E. (1974) The beam on an elastic foundation: matrix solution, *Analytical and Computer Methods in Foundation Engineering*, McGraw-Hill, pp. 147–186.

Ghosh, S.K., Fanella, D.A., Rabbat, B.G. (eds.) (1996) *Notes on ACI 318-95, Building Code Requirements for Structural Concrete with Design Applications*, Portland Cement Association.

Hetenyi, M. (1946) *Beams on Elastic Foundation*, University of Michigan Press, Ann Arbor.

Reese, L.C. (1984) *Handbook on Design of Piles and Drilled Shafts Under Lateral Load*, FHWA-IP-84-11, July, FHWA.

Seely, F.B. and Smith, J.O. (1952) Articles 59-66, Beam on continuous elastic support, *Advanced Mechanics of Materials*, 2nd edn, John Wiley & Sons, Inc., pp. 188–219.

Terzaghi, K. (1955) Evaluation of coefficients of subgrade reaction. *Geotechnique*, **5** (4), 297–326.

Timoshenko, S. and Woinowsky-Krieger, S. (1959) Article 8, Cylindrical bending of a plate on an elastic foundation, *Theory of Plates and Shells*, 2nd edn, McGraw-Hill, pp. 30–32.

17

Footings and Mat Foundations

17.1 Mat Foundations

Mats find applications for foundations with the following conditions:

- Weak soil.
- Area of the spread footings >50% of the total area – if loads are uneven this rule may result in an uneconomical design.
- Soils having variable compressibility – reduce differential settlement by bridging weak and strong soil with the mat that will yield less settlement.
- Reducing hydrostatic uplift by using slab weight.

The following three design conditions, all related to deformation, are involved:

Bearing capacity
Total and differential settlement
Structural integrity and moment capacity of the mat.

Bearing capacity is usually not a problem for mats unless they are resting on cohesive soil for which there is no ϕ since in the Terzaghi bearing capacity equation,

$$q = cN_c + \gamma DN_q + \gamma B/2N_\gamma$$

B is very large and controls. Settlement is governed by the size of the pressure bulb's zone of influence.

Solutions for Soil and Structural Systems using Excel and VBA Programs, First Edition. Robert L. Sogge.
© 2012 John Wiley & Sons, Ltd. Published 2012 by John Wiley & Sons, Ltd.

17.2 Slab Section Stiffness and Moment Capacity

As discussed in Chapter 8, the reduced stiffness of a concrete section that occurs after cracking begins can be derived using:

- Ghosh *et al.* (1996) equations for I_{cr} in Table 8.2, page 8-4 of that text (used in worksheet of Chapter 8).
- ACI (1989) or AASHTO (2010) equations for $EI = (E_cI_g)/5$ or 2.5.
- Program PMEIX solution for reduce EI developed by Reese (1984).

Cracked-section moments of inertia shown in Table 17.1 are determined using the equation for I_{cr} given in Table 8.2 in PCA (1996). These numbers are closer to $EI_g/5$ than $EI_g/2.5$. The design bending moment capacities are determined by program **Beam-LFD** presented previously in Chapter 8.

17.3 Soil–Structure Interaction

The moments in the structure and pressures in the soil are dependent on a non-dimensional ratio

$$S = \sqrt[4]{BkL^4/(4EI)}$$

Table 17.1 Slab design comparison.

Thickness	Reinforcement				I_g	I_{cr}	I_{cr}/I_g	M_d
(in)	Bar size #	Spacing (in)	Area (in²/ ft)	% A_g	(in⁴/ ft)	(in⁴/ ft)		(k-in)
5.5	4	12	0.196	0.30	166	6	0.04	1.8
5.5	4	18	0.131	0.20	166	4	0.02	1.2
5.5	4	24	0.098	0.15	166	3	0.02	1
7	4	12	0.196	0.23	343	18	0.05	3.1
7	4	18	0.131	0.16	343	13	0.04	2.1
7	4	24	0.098	0.12	343	10	0.03	1.6
10	4	12	0.196	0.16	1 000	63	0.06	5.8
10	4	18	0.131	0.11	1 000	44	0.04	3.9
10	4	24	0.098	0.08	1 000	34	0.03	2.9

The following assumptions are involved in the construction of this table:

f_c' (psi)	E (ksi)	f_y (ksi)	Bottom clearance (in)
3 000	3122	60	3

ACI 318-89 Building Code, Section 7.12, $A_s = 0.18\%$ A_g with maximum 18 in spacing.
ACI 318-89 Building Code, Section 10.11.5.2, Equation (10-10) $EI/E_cI_g = 1/5$ to $1/10$,
Equation (10-11) $EI/E_cE_g = 1/2.5$ to $1/5$.
I_{cr} is computed based on PCA (1996) equation in Table 8.2.

which is the characteristic of the governing fourth-order differential equation developed by Hetenyi (1946) (see Chapter 16), where

soil stiffness $= Bk$
structural stiffness $= 4EI/L^4$.

The following ranges of S describe how the slab will operate:

S < π/4 rigid slab or very weak (soft) soil – for which moments, displacements, and soil pressure distributions can be computed based on statics.
π/4 < S < π intermediate soil and structural stiffness – for which the interaction present between the soil and structure can best be determined using a beam–bar finite element model.
S>π flexible slab or very stiff soil – for which soil pressures equal the loading pressures applied.

Typically for reinforced residential slabs, the soil–structure stiffness ratio exceeds π and the slab behaves in a flexible manner.

17.4 Practical Considerations Regarding Slab Reinforcement

Good references on slab design are provided by Kiamco (1997), Ringo and Anderson (1992), and Ulrich (1991).

17.4.1 Advantages of Steel Reinforced Slabs

Advantages of using reinforcement in slabs-on-grade exist in the following situations (CRSI and WRI, 1991):

- **Eliminate joints.** By using a reinforced slab the need for joints and the problems associated with them is eliminated. Slabs can easily extend up to 75 ft in a direction without a control, expansion, or construction joint. Construction joints may be spaced according to the planned size of a single day's pour.
- **Subgrade drag forces.** A reinforced slab can resist forces in the soil due to deep-seated regional movement, perhaps expansion, that results in subgrade drag forces acting on and producing tension in the slab.
- **Cracking.** Although reinforcing steel will not prevent cracking due to shrinkage it will hold the cracks to a tight (hairline) width and thus maximize aggregate interlock load transfer.
- **Curling resistance.** Reinforcing steel placed in the upper half of a concrete slab-on-grade will restrain shrinkage and thus reduce curling.
- **Structural strength.** When cracking occurs, the steel will provide moment capacity for the section.

- **Soft spots in subgrade.** Soft spots created by utility excavation can be spanned by a structurally reinforced slab.
- **Minimizing slab thickness.** Thickening the edges and interior load points can be minimized by the added strength imparted by the reinforcing. By placing steel through the entire slab section, that portion of the slab away from the footing acts as a foundation in distributing the load.
- **Doubtful subgrade.** This is for situations where doubt exists concerning the subgrade support.

17.4.2 Sizing Steel for Bending Moment Capacity

It has been assumed in the past that house slabs will remain uncracked due to the superimposed loadings of the wood frame or masonry structure above. Moments develop in the slab predominantly due to ground movement in many areas. Structurally active steel is required to develop the section's moment capacity. A reinforced slab-on-grade becomes structurally active when the moment imposed on it exceeds the uncracked moment capacity of an unreinforced slab section. A thick slab without much reinforcement can actually cause some problems by its weight. In the case of a soil settling under the edge of a slab, the weight of a thick slab can increase the moments in the slab by 10–15% above thinner slab sections.

Grade 60 steel is used to minimize the amount of steel. Grade 40 steel has the advantage that it will keep cracks somewhat tighter since it is designed for lower strength, thus lowering the deformation at design-level loads. Rebar spacing must be structurally stiff to support workers placing concrete or widely spaced so workers can step between the bars. It is for the support and spacing reasons, among others, that #3 bars are not used. At short spacing they do not provide the adequate rigidity to support workers and at large spacing they do not meet the area of steel reinforcement requirements. For these reasons, grade 40 steel is typically not specified in either structural slabs or slabs-on-grade.

17.4.3 Sizing Steel for Temperature and Shrinkage and Subgrade Movement

The requirements for temperature and shrinkage reinforcement as specified in the ACI 318-89 Building Code Section 7.12 for structural slabs (slabs-on-grade are excluded) state that the minimum steel reinforcement placed perpendicular to the flexural reinforcement is 0.0018 times the gross area of the member for grade 60 steel. The spacing of such reinforcement is a maximum of 18 in. Steel to control shrinkage should be placed at or above mid-depth of the slab.

If the subgrade moves a frictional shear force is built up along the slab. This force creates a tension force that is maximum in the center of the slab. In order to size the steel reinforcement for this condition, the following subgrade drag equation is used:

$$A_s = FLw/(2f_s)$$

where

A_s = area of slab steel (in^2/ft) required each way

F = subgrade friction factor presented in Table 17.2 (PTI, 1980)

L = slab length between joints (ft)

w = slab weight per square foot (12.5 psf/in of thickness)

f_s = working stress of steel = $\frac{2}{3}f_y$ (psi).

For a 7 in thick slab of 100 ft length resting on a soil having a friction factor of 1, the subgrade drag equation yields #4 @ 18 in on center (o.c.) each way or 0.15% (0.0015) of the concrete gross cross-sectional area. This equation shows that increasing the joint spacing increases the amount of steel required. This steel is a maximum at the center of the slab where the frictional shear forces are greatest.

17.4.4 Conclusions on Steel Reinforcement

The quantity of steel reinforcement in a slab is a function of the moment in the slab and shrinkage steel requirements. From the analysis performed it is seen that only slightly more steel is required to resist moment than is required for shrinkage. Placement locations of the steel for these two design conditions can be in opposite portions of the slabs. As a compromise to resisting stresses and controlling shrinkage, placement is often at mid-depth of the section. Bolster (bar) supports are recommended to support the reinforcing steel at its detailed location.

The size of the rebar required in slab design is considerably more than what has been used in the past by many builders. The steel quantity used now is more on the order of the #3 @ 12 in spacing in a 4 in thick slab shell which pool builders, who cannot tolerate cracks in their products, use. In locations where known soil problems exist, #4 @ 6 in o.c. in a 5.5 in thick slab should be used. Such a design is for concrete having a 3000 psi compressive strength.

17.4.5 Welded Wire Reinforcing

Typically welded wire reinforcing (WWR) has replaced welded wire fabric (WWF), denoted as "fabric," due to the small diameters of the wire comprising it. One of the

Table 17.2 Subgrade friction factor from PTI (1980).

Soil type	First movement	Subsequent movement
Plastic clay	2.1	1.3
Blended, washed sand + gravel	1.9	1.3
Granular sub-base	1.7	0.9
Sand layer sheeting	1	0.8
Polyethylene	0.9	0.5

major reasons for this replacement is the thinking that if concrete cracks, even in the smallest degree, moisture can make it to the rebar. If this reinforcing is as small as a wire it could more readily be compromised by rust and deterioration than could a larger rebar. The only advantage of WWF is the close bar spacing that can sometimes help to minimize cracks.

Rebar mats that are used as slab reinforcement must be properly placed and supported at the design location prior to and during the placement of concrete. Pulling the mesh through the concrete up to its design location is not an acceptable procedure.

17.4.6 Post-tensioned Structural Slabs

Instead of using reinforcing steel to make a structural slab section, post-tensioning strands can be used.

Advantages:

- Allows the full thickness of the slab to be effective in developing the moment of inertia, thus a thinner section can be employed.
- Minimizes any cracking by keeping the entire section in compression.

Disadvantages:

- Specially trained craft workers are required to construct the slab.
- A delay is needed while the concrete attains strength before the slab can be post-tensioned.
- Inspection of the procedure is needed.
- Costs exceed a conventionally reinforced slab.
- By soil–structure interaction principles, since a post-tensioned slab section is stiffer than a conventionally reinforced section, it will receive somewhat more moment from the applied loading.

Common strands used for post-tensioned slabs are $1/2$ in diameter – 270 ksi yield. A spacing of 6 ft o.c. is often used for residential slabs. The slab thickness is often 4 in and the concrete has a minimum strength of 2500 psi. This steel quantity is much less than a conventionally reinforced structural slab. When stressed to 70% of the ultimate strength, the 37 kips applies a pre-stress force of approximately 125 psi to the slab cross-section before losses due to relaxation. Typically, the post-tensioning stress induced in slabs, if it is approximately 50 psi, is minimal.

The moment capacity of such a pre-stress force in a 4 in slab is approximately 0.13 k-ft. Therefore the slab derives much of its strength from the waffle-like grade beams that go through the center and the edge of the slab. The waffle slab configuration handles any moments induced by subgrade movement and the post-tensioning is used to control shrinkage. It is still advantageous to use conventional reinforcement in the short direction to control shrinkage cracks if tendons are not used in that direction. Design of a post-tensioned slab is best accomplished by using the Post-Tensioning

Institute solution approach and employing a computer program to perform the design procedures (PTI, 1980).

17.5 Case Study – House Slab Foundations in Tucson, Arizona

The southwestern United States hosts many engineering challenges that are often unique to the specific environmental regime. In a northwestern portion of Tucson, situated in a regional lowland which is adjacent to the Canada del Oro River, a seasonal dry wash which accumulates approximately 80 square miles of upstream waters, a residential area is developing. Aerial photographs showed most of this area to be an alluvial fill deposit consisting of sands and silts that formed behind check dams installed by the US Army Corps of Engineers in the 1950s. Approximately 5 ft of compacted fill had been placed over the weaker soils.

Heavy rains in 1990 and 1993 brought distress to some of the houses within the aforementioned development. The Department of the Interior's Housing and Urban Development (HUD) became involved in the development when problems were noted with properties having federally insured loans. Strong concerns were placed on four builders in this area to evaluate alternative foundations, and to bring solutions forward to alleviate the distress from the soil movement in the area.

For approximately 15 years it was previously known that soil having problem characteristics that could be treated with site-specific slab-on-grade reinforcement could be identified. For a long period, only one geotechnical consulting firm recommended that reinforcement be used in house slabs-on-grade. HUD also had seen the need for such a requirement and was ready to require some reinforcement in all house slabs, but had delayed such an announcement until May 1993.

Many of the concrete contractors and their structural designers had been "selling" thick slabs having reinforcement which barely developed the cross-sectional stiffness properties of the concrete. This design was based on the concrete section not cracking, and the design was also based on little soil strength parameter input. Also, vendors had been "selling" post-tensioned slabs as a cure-all whether they were needed or not.

Following meetings between the builders and officials of HUD the conclusion was reached that, even though eight soil reports had been done by three different soil firms, geotechnical recommendations for a structural slab solution to the soil movement problems inherent with the site should be developed.

17.6 Example 17.1 House Slab

This example presents the analysis of the house slab for conditions discussed in Section 17.5. Using the **PFrame** analysis program with a Hetenyi beam-on-elastic foundation model, the house floor slabs were analyzed for the loading conditions of wall and slab loads and soil settlement. This slab is to be designed to accommodate the settlement in the area without developing large cracks.

17.6.1 Input Data

The input data for a house with a slab that is loaded at each end by the roof trusses supporting the tile roof and the extent of the soft foundation zone beneath the slab is shown in the **Example 17.1** worksheet.

17.6.2 Results

For the load and support displacement conditions stated, the moment, deflection, and soil pressure distributions can readily be computed in the workbook of **Example 17.1** for slabs having sections 5.5, 7, and 10 in thick.

The structural (ultimate) design load moment capacities and the cross-sectional moment properties are developed for specific rebar locations and concrete strengths and are presented in Table 17.1. As can be seen from this table, the steel quantity for shrinkage and temperature, 0.20% of A_{gross} as recommended by ACI, and approximately 0.15% as determined from the subgrade drag equation, yield considerably greater amounts than required to develop bending moment capacity.

For a 5.5 in thick slab the moment in the section subject to the settlement conditions specified of 1 k-ft is approximately double that due to the loads alone from the structure and #4 @ 24 in o.c. are required. The reduced coefficient of subgrade reaction for the soil at its end is equivalent to imposing a support displacement of approximately 0.5 in that tapers to 0 in 8 ft away. The maximum total slab settlement is 0.7 in or approximately the 0.2 in overall settlement plus the differential settlement. The moments induced in the slab and the slab capacity presented in Table 17.1 show that all slab thicknesses with the proper reinforcement will work. The moment calculation requires that steel be placed in the top portion of the slab.

17.6.3 Conclusions

In other situations where soils with considerable swell potential exist, slabs need to be designed to accommodate such movement. This approach allows a better designed slab that will respond appropriately to the soil conditions present in the field. Conventionally reinforced slabs can provide a cost saving over post-tensioned slabs and provide the margin of safety required.

By applying a structural solution to a soil problem, advantages of the structural slab solution are reduced over excavation and recompaction, the elimination of joints, control of cracking, minimizing slab thickness, and providing strength for curling resistance, bridging soft spots in the subgrade, and subgrade drag forces. Alternatively, post-tensioned structural slabs, where much of the entire thickness of the slab is effective, can be used to minimize cracking.

References

AASHTO LRFD (2010) *Bridge Design Specifications*, 5th edn, American Assocation of State Highway and Transportation Officials.

ACI (1989) *Building Code Requirements for Reinforced Concrete*, ACI 318-9, American Concrete Institute.

CRSI and WRI (1991) Reinforcing Steel in Slabs-on-Grade, Engineering Data Report No. 37, Concrete Reinforcing Steel Institute, Schaumburg, and Wire Reinforcement Institute Tech. Facts TF 701, Washington, DC.

Hetenyi, M. (1946) *Beams on Elastic Foundation*, University of Michigan Press, Ann Arbor.

Kiamco, C. (1997) A structural look at slabs on grade. *Concrete International*, pp. 45–49.

PCA (1988) *MATS, Computer Program for Analysis and Design of Foundation Mats and Combined Footings*, Portland Cement Association, Skokie, IL.

PTI (1980) *Design and Construction of Post-Tensioned Slabs on Ground*, The Post-Tensioning Institute, Phoenix, AZ.

Reese, L.C. (1984) *Handbook on Design of Piles and Drilled Shafts Under Lateral Load*, FHWA-IP-84-11, July, FHWA.

Ringo, B.C. and Anderson, R.B. (1992) *Designing Floor Slabs on Grade*, The Aberdeen Group, Addison, IL.

Ulrich, E.J. Jr. (1991) Subgrade reaction in mat foundation design. *Concrete International*, pp. 41–50.

18

Laterally Loaded Piles

This chapter covers the analysis of laterally loaded piles and piers. Cantilevered piles fall into this category but are covered more extensively in Chapter 19 along with anchored sheet piles. The description of pile refers to driven steel H-piles or pre-stressed concrete or timber piles. The description pier, often denoted as a shaft or caisson, refers to cast-in-place concrete structures. A shell that is driven with a mandrel to later be filled up with concrete is also denoted as a pile.

18.1 Theory – Classical Differential Equation Solution

The differential equation for a beam-on-elastic foundation governs pier behavior (Hetenyi, 1946) as was presented in Chapter 16:

$$EI\frac{d^4y}{dx^4} = -p + q = -Bk_1y + q$$

where

$x =$ distance along pile (in this equation only)
$y =$ pile horizontal deflection (in this equation only)
$p =$ soil pressure $= Bk_hy$
$q =$ applied load/unit length
$B =$ width of beam
$k_1 =$ constant of horizontal (lateral) subgrade reaction for a 1 ft wide strip.

The solution to this equation fulfills the requirements of equilibrium, stress–strain, and geometric compatibility. The characteristic of this differential equation solution is the same non-dimensional constant as for a beam-on-elastic foundation, with a slight change due to the different definition and value of the coefficient of subgrade reaction for buried structures:

$$S = (Bk_{max}L^4/(4EI))^{1/4}$$

Solutions for Soil and Structural Systems using Excel and VBA Programs, First Edition. Robert L. Sogge.
© 2012 John Wiley & Sons, Ltd. Published 2012 by John Wiley & Sons, Ltd.

The soil stiffness is $B(k_1 z/B)$, where k_1 is the constant of horizontal subgrade reaction for a 1 ft wide strip and z is the distance below the ground surface. The structural stiffness for a beam is related to $4EI/L^4$ as shown in Chapter 3.

With these changes laterally loaded piers (also known as piles, shafts, and caissons) can be analyzed using the program **PFrame**, in a manner similar to that used for a beam-on-elastic foundation.

18.2 Conventional Analysis

Analysis of laterally loaded piles including coverage using the Broms' theory (Broms, 1964a, b) that simplified the conventional static approach is given in the book by Poulos and Davis (1980), *Pile Foundation Analysis and Design*, Sections 7.2.1 and 7.2.2, pages 144–152.

18.3 Beam–Bar Finite Element Solution

The finite element solution of a laterally loaded pier system models the soil–structure system by idealizing the structure as beam members and representing the soil by bars or springs (Bowles, 1977; Clough and Tsui, 1977; Desai and Kuppusamy, 1978). The beam–bar finite element model used to simulate the pier–soil system is represented in Figure 18.1. The length of the pier above ground level can be divided into any number

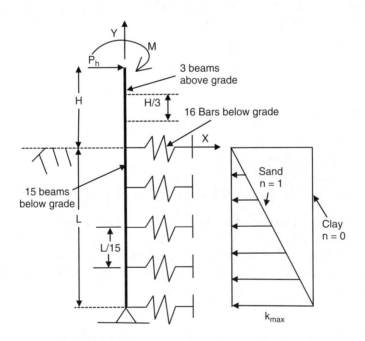

Figure 18.1   Finite element model.

of beam members. The length below ground should be divided into at least 12 but no more than 20 members. On many of the input spreadsheets, 15 beam element divisions are used below grade to represent the embedded portion and 3 elements above grade. The three elements above grade permit the modeling of a cantilevered pile with loads applied above grade.

The VBA program **PFrame** that was introduced in Part Two of this book is used to analyze the beam–bar models. As a general frame program **PFrame** is able to analyze any general soil–structure system modeled using beams and bars. A system consisting of many piers as well as the entire above-ground structural system can readily be incorporated into the model.

18.3.1 Pier Support Conditions

The top of the pier can be supported as shown in Figure 18.2 as:

- free (lateral and moment loads can be applied);
- pinned horizontally (only applied moment loads);
- fixed against rotation (only applied lateral load); or
- a rotational spring (only applied lateral load).

All joints at and below the ground surface including the bottom joint of the pier are supported laterally by the attached spring stiffness. The top joint is always present in the analysis for input of other support conditions, if required.

18.3.2 Pier Loadings

Lateral loads applied at the top end of the structure can consist of (Figure 18.2):

- Lateral force
- Moment
- Both.

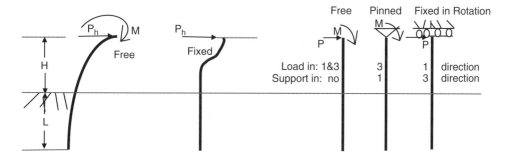

Figure 18.2   Pier supports and load conditions.

If a load is applied in a rotational or translational direction to the top of the pier, that direction cannot be supported or in essence the load will act as an imposed displacement. Any vertical load is essentially decoupled in this analysis and is not shown. It need not be part of the input to a single-pier analysis. The moment and force loadings for the fixed-end condition of no displacement of the joints are zero since the lateral forces on each side of the pier exactly balance each other resulting in no net fixed-end force (FEF). The FEF loading condition is thus different from those for foundation and anchored sheet piles.

18.3.3 Beam Representation of Pier

Pier cross-sections of any circular, square, rectangular, or H-pile shape can be prescribed. Strength properties for reinforced concrete piers are prescribed for a concrete section to represent the state of cracking. Cracked-section properties are simulated by using a reduced elastic modulus value and a cracked-section moment of inertia, I_{cr}.

18.3.4 Spring (Bar) Representation of Soil

The soil media are modeled using the Winkler assumption of independent springs. The soil is represented by bar (spring) elements. The soil strength is specified by k_h, the coefficient of horizontal subgrade reaction that relates the net soil pressure to the deflection.

The force–deformation relations, $F = [S]x$, for a bar member used to represent a soil are developed as follows:

$$F = Ap \quad \text{Equilibrium}$$
$$p = E \quad \text{Stress–Strain}$$
$$\varepsilon = x/L \quad \text{Geometric Compatibility}$$

where

F = force
A = area
p = stress
E = elastic modulus
ε = strain
x = displacement
L = length of the member.

Combining the above relations for the bar yields

$$F = (AE/L)x$$

For soil this bar element represents the soil's uniaxial resistance

$$p = k_h x \quad \text{or} \quad F = (Ak_h)x$$

where k_h = the coefficient of horizontal subgrade reaction.

Therefore, to use a bar element to simulate a soil, let A equal the contributory soil area, which is the contact area of the soil adjacent to the pier, or B times a length equal to the sum of half the distance on each side of the joint connecting the bar and beam members:

$$E = k_h \text{ of soil}$$

$$L = \text{unity}$$

The spring (bar) force divided by the contributory area of the spring yields the soil pressure.

The finite element idealization of a spring is obtained using beam members having no bending stiffness (moment of inertia $I = 0$) and all degrees of freedom at their support end supported.

The bar elements, with Winkler-type spring resistances that are independent of each other, produce stiffnesses only on the diagonal of the stiffness array. The constitutive equations for this model contain coefficients formed by the addition of the beam element and bar element stiffnesses.

18.3.5 Input Data Workbook for General Configurations

For certain problems having the same geometric configuration such as a vertical line, an Excel worksheet can be developed to automatically generate data for input to

---

 Example 18.1   ADOT lateral load test.

ADOT LatLdTest   I-10 Advance Test Pile Program - W. Papago/I-10 Inner Loop HNTB 6/94

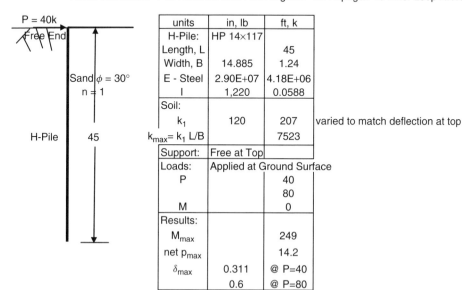

	units	in, lb	ft, k	
H-Pile:	HP 14×117			
Length, L			45	
Width, B		14.885	1.24	
E - Steel		2.90E+07	4.18E+06	
I		1,220	0.0588	
Soil:				
k_1		120	207	varied to match deflection at top
$k_{max} = k_1 \, L/B$			7523	
Support:	Free at Top			
Loads:	Applied at Ground Surface			
P			40	
			80	
M			0	
Results:				
M_{max}			249	
net p_{max}			14.2	
δ_{max}		0.311	@ P=40	
		0.6	@ P=80	

P = 40k
Free End

Sand $\phi = 30°$
n = 1

H-Pile   45

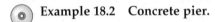

Example 18.2   Concrete pier.

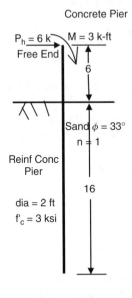

Concrete Pier

units	in, k	ft, k
Conc Pier		
Length, L		16
Dia, B		2
E	f'c = 3 ksi	4.05E+05
I	0	0.349
Soil:		
k_1		30
$k_{max}= k_1 L/B$		240
Support:	Fixed at Top	
Loads:	Applied at Top	
P		6
M		3
Results:		
M_{max}		55.5
P_{max}		0.700
δ_{max}	0.80	0.067

the program **PFrame**. For such geometries, data for cell input consists of the pier height above and below the ground surface, support conditions on the top of the pier (free, pinned, or fixed), applied lateral load and moment (if free) at the top of the pier, applied lateral load (if fixed), applied moment (if pinned) at the top, pier cross-sectional dimensions and concrete strength, and soil friction angle (sand) or unconfined compressive strength (clay), and parameters which define strength variation with depth. Input data for many of **Examples 18.1–18.7** is developed by this generalized method in their associated worksheets.

On the worksheet the output units will match the selected input units; kips and ft should normally be used. The data generation worksheet computes the joint coordinates by dividing the pier into 3 beam elements above ground level and 15 beam elements below ground level. The soil is represented by 16 spring elements. The top pier element joint, if supported in rotation and translation, is given a fixed support condition. Coefficients of horizontal subgrade reaction are developed from soil ϕ angles and unconfined compressive strengths and automatically assigned to the members representing the soil.

Output, consisting of the moments, displacements, and soil pressures, is sent to the worksheet by the VBA macro program. This output is charted in a display that shows their values along the length of the pier. An evaluation of the acceptability of the moments and soil pressures can be had by a comparison to charted lines showing the approximate allowable values.

Example 18.3   Timber pile. Design Example from Sogge (1981, pp. 1191–1195).

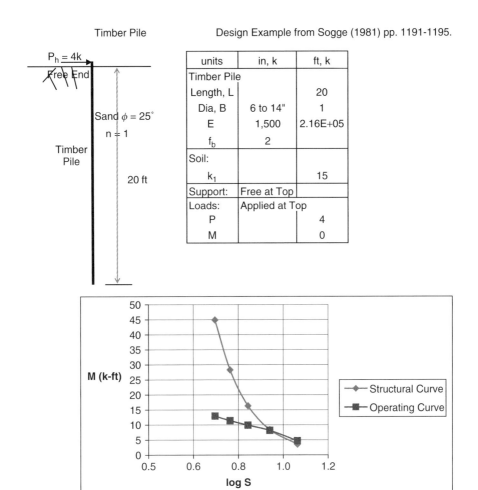

units	in, k	ft, k
Timber Pile		
Length, L		20
Dia, B	6 to 14"	1
E	1,500	2.16E+05
f_b	2	
Soil:		
k_1		15
Support:	Free at Top	
Loads:	Applied at Top	
P		4
M		0

Timber Pile

Design Example from Sogge (1981) pp. 1191-1195.

$P_h = 4k$

Free End

Sand $\phi = 25°$

n = 1

Timber Pile

20 ft

18.4 Structural Stiffness

With regard to drilled shafts, Reese (1984) stated that only a short portion of a shaft will crack at a large developed moment. Thus using an *EI* for a section (obtained using his program PMEIX), when the developed moment was 13% of the ultimate moment, was proper for refining deflection computations. At this strain level $EI_{cr}/(EI)_g$ is approximately 1/2.

Example 18.4   Concrete Pier 2 soils.

Concrete Pier 2 soils

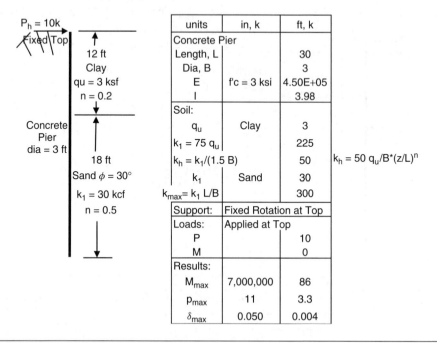

units		in, k	ft, k
Concrete Pier			
Length, L			30
Dia, B			3
E	f'c = 3 ksi	4.50E+05	
I			3.98
Soil:			
q_u	Clay		3
$k_1 = 75\, q_u$			225
$k_h = k_1/(1.5\ B)$			50
k_1	Sand		30
$k_{max} = k_1\ L/B$			300
Support:	Fixed Rotation at Top		
Loads:	Applied at Top		
P			10
M			0
Results:			
M_{max}		7,000,000	86
P_{max}		11	3.3
δ_{max}		0.050	0.004

Left figure labels:
$P_h = 10k$
Fixed Top
12 ft
Clay
qu = 3 ksf
n = 0.2

Concrete Pier dia = 3 ft
18 ft
Sand $\phi = 30°$
$k_1 = 30$ kcf
n = 0.5

$k_h = 50\ q_u/B^*(z/L)^n$

The stiffness of the member section, $EI = M/\phi$, as defined by elastic modulus times the moment of inertia, decreases as the moment in and the rotation of the section increase and the section cracks. Since soil–structure interaction is governed by the structural stiffness, the proper cracked-section stiffness, I_{cr}, should be used for the member. It is the change in cross-sectional moment of inertia properties relative to other members framing into a joint that governs the distribution of moments in that joint. If all members change similarly one could use the gross cross-sectional properties for all members.

The cracked-section and strength properties can be calculated by using one of the three methods discussed in Chapter 8 for computing the strength and section properties of a cracked concrete section:

- An approximate evaluation of slenderness effects and cracked-section properties is addressed in the ACI (1989) Building Code, Article 10.11.5.2, Moment Magnification, and the AASHTO (2010) code, Article 5.7.4.3, Approximate Evaluation of Slenderness Effects. The cracked-section properties are simulated by using a reduced elastic modulus value and a cracked-section moment of inertia, I. The combined reduction

representing the slenderness of the section is taken as $EI_g/2.5$. This reduction is accomplished by reducing the elastic modulus by 10% and using $0.444I_g$.
- Using program PMEIX (Reese, 1984; Reese and Allen, 1977).
- Using the cracked-section modulus presented in Ghosh *et al.* (1996).

18.5 Soil Strength

The strength of the soil as represented by the discrete bars (springs) is specified by the coefficient of horizontal subgrade reaction, k_h, which relates the soil pressure to the deformation of the structure in the soil. This coefficient for sands varies with confining pressure in a manner that is depth dependent. Its variation with depth, or confining pressure, may be constant or linear or some variation in between. Values of k_h decrease for increasing structure widths due to the increased size of the pressure bulb (Terzaghi, 1955).

Example 18.5   (a) Bridge bent – single pier. Single pier selected from Example 18.6 Bridge bent. (b) DoubleSpring. (c) DoubleSpring – imposed displacement.

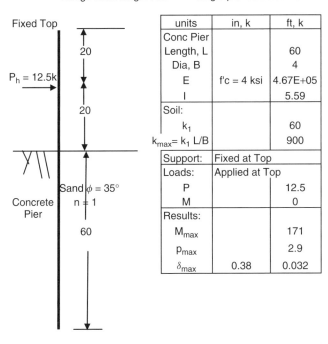

Bridge Bent-Single Pier   Single pier selected from Example 18.6 Bridge Bent

units	in, k	ft, k
Conc Pier		
Length, L		60
Dia, B		4
E	f'c = 4 ksi	4.67E+05
I		5.59
Soil:		
k_1		60
$k_{max}= k_1 L/B$		900
Support:	Fixed at Top	
Loads:	Applied at Top	
P		12.5
M		0
Results:		
M_{max}		171
p_{max}		2.9
δ_{max}	0.38	0.032

Fixed Top

20

$P_h = 12.5k$

20

Sand $\phi = 35°$
n ≠ 1

Concrete
Pier

60

Example 18.6 BridgeBentwithPiers.

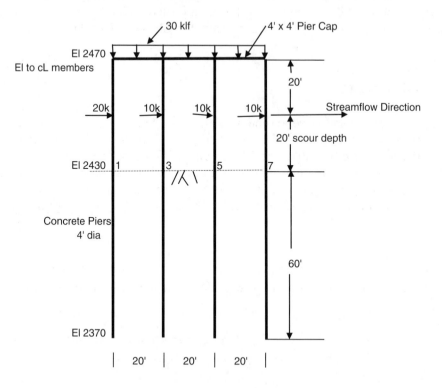

Vertical Load due to traffic
Lateral (transverse) Load due to Streamflow against Piers
Fixed Support condition taken at El 2430 where M_{max} occurs in Pier

units	in, k	ft, k
Results:		
M_{max}		174
p_{max}		0.4
δ_{max}	0.40	0.033

An expression for k_h at any depth, y, below the ground surface that incorporates a linear variation with depth (confining pressure) and the pier width or diameter, B, is given by the equation

$$k_h = k_1 y/B$$

in which k_1 is the constant of horizontal subgrade reaction for a 1 ft wide strip that relates the net soil pressure to the deflection. The distance y term could easily have been defined as z for these buried-type soil stiffnesses as will sometimes be done. This coefficient

Example 18.7 (a) **Concrete Pier H = 0 Solution using PFrame macro. (b) Concrete Pier H = 0 Solution using Excel Workbook only (like Example 18.5 with H = 0).**

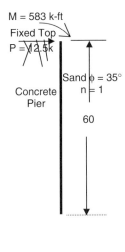

units	in, k	ft, k
Conc Pier		
Length, L		60
Dia, B		4
E	f'_c = 4 ksi	4.67E+05
I		5.59
Soil:		
k_1		60
$k_{max}= k_1 L/B$		900
Support:	Fixed at Top	
Loads:	Applied at Top	
P		12.5
M		0
Results:		
M_{max}		97
P_{max}		0.2
δ_{max}	0.38	0.032

supplies the net resistance of the soil on both sides of the structure. This net force is the resisting passive minus active-type stress existing on opposite sides of the pile.

The units of k_h and k_1 are the same, kcf. The maximum value of k_h denoted by k_{max} occurs when $z = L$, the length of the embedded portion of the pier, yielding

$$k_{max} = k_1 L/B$$

Different variations of k_h with depth, y, can be described by using a shape exponent, n, in the following relation:

$$k_h = k_{max}(y/L)^n$$

For vertical piers in sands the soil stiffness defined by the coefficient of horizontal subgrade reaction approximately varies with depth in a linear manner in accordance with strength increases which sands display with confining pressure (i.e., $n = 1$). For cohesive soils, particularly overconsolidated clays, k_h is not very confining pressure dependent (i.e., $n = 0$). Values for n are presented in Figure 18.3.

18.5.1 Coefficient of Horizontal Subgrade Reaction Values

Values of the coefficient of horizontal subgrade reaction, k_h, and the associated constant of subgrade reaction, k_1, on which it is based are not determined directly in a laboratory

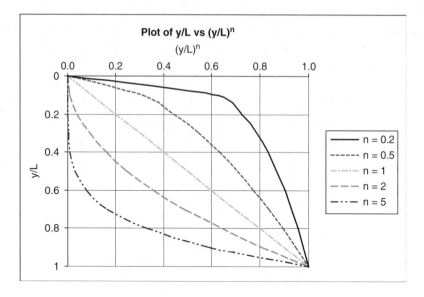

Figure 18.3 Plot of y/L versus $(y/L)^n$.

test. Instead, they are usually derived from the measurement of load, deflection, and slope on instrumented piles in a lateral load test conducted in the field. Generally, the piers need only be instrumented to a depth of 10 times the pier diameter. The derivation of k_h consists of matching measured slope and deflection values with those computed from an analysis of a pile subject to the applied load by varying k_h values. An example showing this back calculation of a k_h value from lateral load test data is presented in **Example 18.1**.

Another method of determining k_h values is by relating them to soil strength properties such as E that can be determined in a laboratory. Values of k_1, the constant of horizontal subgrade reaction, are given in Table 10.1 on soil strength properties in Chapter 10.

18.5.1.1 Sands (Cohesionless Soils)

Values of k_1, representing the stiffnesses of the soil in the combined active and passive state existing on each side of the pier, are presented in Table 10.1 for wet and submerged sands (Habibaghi and Langer, 1983). A relation between ϕ and k_1 can be developed by fitting a curve to the data points relating both parameters. This curve fitting is described in Chapter 13. For sands, the values of k_1, and the unit weight, γ, can be automatically calculated based on the input friction angle, ϕ, using the following relations:

$$k_1 = 4.1043 \exp\{(\ln(\phi) - 2.4524)^2/0.4531\}$$

$$\gamma(\text{pcf}) = 1/(0.0330 - 0.0068 \ln(\phi))$$

These equations have been developed using the program **CurveFit** (Cox, 1992) that is a VBA macro attached to an Excel workbook. This **CurveFit** program provides more equations than those in the "add a trendline" feature of Excel charts.

18.5.1.2 Clays (Cohesive Soils)

For clays, the following relationship presented for the lateral k, or k_h, was given by Terzaghi (1955):

$$k_h = k_1/(1.5B) = 75q_u/(1.5B) = 50q_u/B(\text{kcf})$$

in which q_u is the unconfined or unconsolidated–undrained (UU) compressive strength, and k_1 is a constant of subgrade reaction. Adding the shape factor relation $(y/L)^n$ to prescribe variations of k_h with depth yields

$$k_h = 50q_u/B(y/L)^n$$

For vertical piles in stiff clays, the values of k_1, representing the stiffnesses of the soil in the combined active and passive state existing on each side of the pier, are approximately equal to the k_1 values given for vertical loadings applied to beams resting on a clay surface. Representative values of k_h for clay materials are presented in Table 10.1. The value of the horizontal subgrade reaction coefficient for the spring element at ground surface can be approximated respectively as $1/4$ and $1/2$ of the next lowest spring value for sand and clay soils that display a variation with depth.

Piers in non-homogeneous (i.e., layered) soil systems can be approximated by specifying a different variation in depth using the n coefficient to represent the different material properties of the bar elements along the pier length. To be more precise, different soil types can be specified at various depths. This latter refinement is rarely needed as different properties with depth can be approximated sufficiently using a variation prescribed by the n coefficient in $(y/L)^n$.

18.6 Soil–Structure Interaction

The same characteristic of the differential equation for a beam-on-elastic foundation governs pier behavior. The non-dimensional constant, S, is a ratio of the soil stiffness to the structural stiffness:

$$S = (Bk_{max}L^4/(4EI))^{1/4}$$

The structural stiffness is proportional to $(4EI/L^4)^{1/4}$.

The soil stiffness is proportional to $(Bk)^{1/4}$. For pier systems $k_{max} = k_1L/B$ is substituted for k.

The response parameters such as moment in the pier, soil pressure, and deflection are dependent on the ratio, S, or the soil–structure stiffness ratio. Soil–structure interaction principles are shown in Figure 18.4. For the lateral load, Q, it is evident that weaker

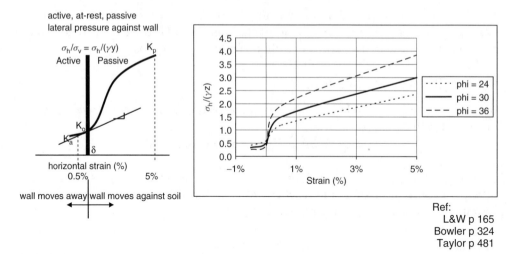

Figure 18.4 Soil pressure–displacement (active, at-rest, and passive lateral pressure against wall).

soils as indicated by smaller S values cause a pier to act stiffer and receive more load (thus moment). Conversely, stronger soils, as indicated by larger S values, cause less moment to be carried by the pier. In other words, very flexible piers or piers in stiff soil support, as indicated by large S, develop less moment than rigid piers or piers in weak soil support.

The behavior of piers can be classified according to their relative stiffness, as delineated by the non-dimensional log S ratio, into the following approximate groupings:

$$\log S < 0.3 \quad 0.3 < \log S < 1.3 \quad \log S > 1.3$$
$$\text{Rigid} \qquad \text{Intermediate} \qquad \text{Flexible}$$

Interaction principles are evident in that

Structure: Stiff or rigid		large moment in structure
or	yields	and
Soil: Weak or soft		low soil pressure

Structure: Flexible		small moment in structure
or	yields	and
Soil: Strong or stiff		high soil pressure

The computed deflections are approximately the same for a rigid or flexible pile.

A pile can be made to behave in a relatively rigid manner by either being extremely stiff itself or being embedded in an extremely weak soil. The more rigid a pile is, the greater the M in its section at greater distances from the top of the pile. For such a pile, the soil response, as measured by p_{max}, is small and occurs at deeper depths.

A comparison of the moment induced in the structure, denoted as the SSI curve, from the loading and the moment capacity of the structure, denoted as the Structural Capacity curve, for a timber pile is presented as **Example 18.3 (Timber Pile)**.

It is necessary to design a structure whose moment capacity as derived from the structural capacity curves is greater than the moment developed in the structure from the loading.

Soil–structure interaction does not enter into structures that are not supported or loaded by soil. If they are loaded by soil, interaction may or may not be present depending on how the soil load is applied to the structure. The frame examples presented in Chapter 6 do not result in any interaction between the soil support and the structure since the support is fixed and has infinite stiffness by this support indication. Therefore it is not necessary to define the stiffness of the structure to a high degree. Typically properties of stiffness, E, A, and I, are based on gross uncracked sections for all members. If a more accurate cracked-section modulus were determined for each of the members, little change would result since their ratio would be essentially the same.

On the other hand a laterally loaded pier presents a very good example of a soil–structure interaction system. For such a system it is necessary to more precisely define the stiffness of the structure since the ratio of its value to that of the soil will determine the structural response. Thus E, A, and I based on gross uncracked sections should be computed.

18.7 Soil Pressures on Each Side of Pier

In determining the portion of the resistance that is due to passive resistance and active pressure it is beneficial to think in terms of a model having a spring on each side of the pile representing the soil on each side. The easiest way to visualize what is going on with the soil is to envision the springs on each side of the pile being compressed by the K_o pressure present before deformations occur. Then, as movement occurs, the active side pressure is reduced from K_o to near K_a and the passive side pressure is increased above K_o to K_p (see Figure 18.4). On the active side there exists no resistance to movement as the pile moves away from the soil. There is a constant pressure on this active side yielding no resistance, tensile force, but always a compressive force.

The soil pressures and the soil resistances on the active and passive sides of the pile are related to the active, at-rest, and passive earth pressure coefficients of approximately $1/3$, $1/2$, and 3 and the deformations that occur in the active and passive states of approximately $1/2$ and 5%. This relation is shown graphically in Figure 18.4. From this relation E and therefore k values of $3.6\,z$ and $5.8\,z$ were developed for the active and passive cases.

For large deformations that occur especially near the top of the pile, the change in pressure is predominantly due to the pressure that develops on the passive side. By just comparing the coefficients of earth pressure in the passive and active states, K_p/K_a, we see that the ratio is approximately 10. Even if a FEF is added to the computed net pressure it has little influence on the resultant pressure on a side. Therefore the net resistance, as represented by k_h, mainly represents the passive side's resistance to movement and the development of pressure on that side.

For low load levels resulting in low displacements, a tangent to the curve through the K_o pressure state shown in the Figure 18.4 representation of the stiffness of the soil in the active and passive states can be used. This model quickly breaks down if the loads are large and the passive resistance dominates the active pressure exerted. If loads are small resulting in small displacements, or if stiffnesses are selected properly, this tensile force should be more than cancelled when added to the K_o FEF of the initial K_o overburden force.

It is not possible to have a spring representing the active state and another representing the passive state as these states and their locations are not known a priori, before the displacements are determined. Only at that stage of analysis are the active and passive sides of the pile known.

18.7.1 Allocation of Soil Resistance to Passive and Active Soil States

The net soil resistance cannot be divided into its passive and active components before the displacement calculation is done, because it is not known what state the soil on each side of the pile is in. From an analysis using the net stiffness or resistance on a joint, the displacements are determined and the net resisting pressure can be computed. This net resistance offered by the soil to the pile movement comprises a passive resistance as the pile goes into the soil and increases its pressure from K_o to K_p minus an active relieving or released pressure as the soil pressure on the active side of the pile moves away from the pile and the pressure decreases, or is relieved from K_o to K_a. At this stage, when the displacements are known, the true active and passive side pressures in the springs can be determined.

For a positive P_x force, the passive and active sides can be determined from the computed displacement pattern as follows:

$+P_x$	δ	Passive	Active
+	Right	Left
−	Left	Right

The net resistance, or lack of resistance in the case of the active side pressure, can now be allocated to each side. That is, 80% passive, 20% active. Since the deformation is equal on the passive and active sides of the pile, this statement in essence is saying that the passive side spring stiffness is four times the active side spring stiffness.

After the displacements and forces (and thus pressures) in each of the one-spring soil elements due to the applied lateral load, P, at the top of the pier are computed, the soil pressure on each of the active and passive sides of a pile can be determined as follows:

- On the passive side (deformation of pier into soil): add (%Passive) $k_h(-\delta)$ to $K_o \gamma y$.

- On the active side (deformation of pier away from soil): add (%Active) $k_h(-\delta)$ to $K_o \gamma y$.

Here, the initial soil pressure on each side of the pier is equal to the overburden pressure, $K_o \gamma y$.

The above approach is equivalent to using the following FEF approach. The initial lateral soil pressure on each side of the pile is the K_o condition due to the overburden soil pressure. This soil pressure could be applied as FEFs to springs on each side of the pier, but since the forces due to the initial overburden pressures balance each other, no net force needs to be applied to the adjoining joint.

Even if one spring element is used, no FEF is applied as the initial lateral soil pressures on each side are considered balanced at the joint where the spring element attaches to the pile and where it is supported. After the forces in the springs due to the lateral load P are determined, the FEFs need to be added to the computed element forces (stresses) in the spring on each side of the pile.

A fixed-end moment (FEM) input, applied automatically through a distributed load approach to the pier model joints, is not required since the FEM on the beam would be balanced from the equal at-rest initial pressure on each side of the beam (pier).

18.7.2 Representing Soil Resistance by the One-Spring Model

The net resistance is represented by the k_h of the one spring at each depth. After the displacements are calculated using this model, an allocation of resistance can be done as mentioned previously where the K_o FEFs are added to a percentage of the determined spring force based on the soil state being active or passive. Since the soil resistance is given as a net resistance for the soil on both sides of the pile, the one-spring model representing this resistance is best.

18.7.3 Representing Soil Resistance by the Two-Spring Model

A two-spring model really offers nothing more than an aid to visualizing the stress state on both sides of the pile and introducing some topics on imposed support displacements. Such a model is really only useful at low load levels that result in low displacements. For such a displacement state a tangent to the curve through the K_o pressure state shown in Figure 18.4 is an adequate representation of the stiffness of the soil. Then equal resisting stiffnesses representing the active and passive states, where each spring has half the stiffness of a one-spring model, can be used. For the two-spring model the spring force is calculated directly and does not have to be halved as in the case of the one-spring model. With such a model the FEFs must be added to the computed spring forces resulting from the lateral load, P.

To assign specific stiffnesses to the two springs of a two-spring model representing the active state and another representing the passive state would require an iterative solution fitting the different spring stiffnesses to the displacement state existing on each side of the pier.

18.7.3.1 Using FEFs

A two-spring model presents another way of envisioning what is going on with the soil. The difficult part to understand is that on the active side of a two-spring model, as the soil is unloading, what is the "resistance" that it is offering to any wall movement? It can be confusing since a tensile force exists in the active side spring. How can a soil in tension offer resistance?

This tension state is not what is happening in the field. It is not until the FEFs are added to the spring forces on each side that some clarity or an element of sense comes to the model. A non-resisting pressure that is decreasing from K_o by the amount of the tensile force determined for the spring results on the active side. Resistance is increased on the passive side by this same K_o compressive force, yielding the same net resistance.

What is seen is that the FEFs or the forces resulting from an imposed K_o deformation balance each other leaving a spring in compression resistance and another in tension resistance as it is unloaded. The use of FEFs only enters when computing the forces in the soil on each side, and for this computation two springs were not needed.

An example of a pier bent using a two-spring model to represent the soil on each side of the pier is provided as **Example 18.5b**. Note that for such a model the k_h value is divided by 2 to represent the stiffness on each side of the pier. When the FEF (pressure) of $K_o \gamma y$ is added to each side the resultant active and passive pressure results.

18.7.3.2 Using Imposed Support Displacements

A two-spring model can be employed in a different manner. In this two-spring model representation the supports are "loaded" with a deflection equal to the initial lateral soil pressure, K_o, condition of the overburden soil pressure acting to compress the springs. The imposed support displacement that precompresses the springs on each side to a displacement equivalent to the initial K_o pressure is $\delta = K_o \gamma y A_{contrib}/k$.

In this scenario, the soil acts like a precompressed bar, still offering a force when unloading, though less than when loading. It is pushing, but less so, as the wall moves away from the soil, and the soil pressure decreases from the K_o to the K_a state. Conversely, on the passive side, the resistance has been increased by this K_o state amount, so the net soil resistance remains the same. No FEFs are required to be added in this type of model analysis.

This imposed K_o displacement approach will become very useful later when arches are being analyzed in Chapter 20. In the arch case there will be soil only on one side of the structure. By using this approach the soil pressure is calculated directly and there is no need to add FEFs.

Example 18.5c provides a double-spring model with imposed displacements. The same soil pressures result from using either a one- or two-spring system to model the soil.

In summary there is really no need for a two-spring model for the following reasons:

- The soil stiffness is given in terms of one net resistance.
- A two-spring model requires that the division of stiffness in the soil on each side of the pile be allocated to the active and passive contributions before the displacement calculation is made. This division of net resistance contributions is not always possible.
- A two-spring model only makes sense if the resistance, k, is contributed equally from both sides, that is, the tangent stiffness through the K_o point on the resistance curve.
- For large displacements a 50/50 allocation makes no sense as it can result in larger changes in active side forces than $K_o - K_a$.
- A one-spring model can adequately represent the net resistance provided by the soil and the allocation of pressures to the passive and active sides can be performed after the analysis that determines the displacements and net pressures. This allocation is simpler to do with the one-spring model.

18.8 Limitations of a Beam – Bar Analysis

These are as follows:

- A beam–bar analysis has the limitation that the beam–spring model does not directly relate vertical loading to lateral soil pressures. Therefore, any vertical arching arising from horizontal pier movement is not computed. Likewise, the influence of depth and confining pressure is handled by the linear increase in soil depth and by the dependency modifier variation $k_h = k_1(y/L)^n$.
- The approach taken assumes that the soil properties for soil reactions that are less than 1/3 to 1/2 of the ultimate reaction can be sufficiently approximated by a linear initial tangent or secant relationship between pressure and displacement on both the active and passive sides of the pier. Systems loading the soil into its yield state would require an incremental or iterative analysis using tangent values of k_h that are dependent on the stress level and on the stress state being either active or passive. This PFrame analysis program applies the load in one direction. It does not perform an iterative analysis. To improve on the present analysis, the true p–δ curve should be approximated incrementally or iteratively (Halliburton, 1968).
- An incremental construction sequence is not modeled by this approach.

18.9 Design Procedure

As PFrame is an analysis program and not strong on design options, a design sequence must be carried out by repeated analyses of various analysis options. The following steps should be employed in the moment design of a laterally loaded pile; **Example 18.3** shows each of these design steps for a timber pile (Sogge, 1981, 1984, 1992):

- Select a trial design pile having a specific length, L, elastic modulus, E, and allowable stress, σ_a.
- For the soil properties, determine the coefficient of subgrade reaction, k_1, its maximum value, k_{max}, and its variation with depth, n.
- Calculate the allowable moment, M_a, for piles of various cross-sections having specific L, E, and σ_a properties. Plot M_a versus log S to develop the "structural capacity curve" for this material. The allowable moment can be calculated from the equation $M_a = \sigma_a I / y$, where $y =$ the distance from the centroid of the section to the extreme outer fiber. The moment of inertia I should be based on the gross section. All pile capacity calculations should be based on the soil stiffness property present around the pile. To develop the moment capacity of this pile under the applied axial load and moment, a P–M interaction diagram developed for circular beam–columns is required. An ultimate or strength capacity for the beam–column can be developed using program PMEIX (1988) that was described in Chapter 8.
- By running the program for a variety of values for $S = (Bk_{max}L^4/(4EI))^{1/4}$, develop an interaction curve that plots the relation between the M_{max} induced in the pile by the applied loads.
- Plot M_{max} versus log S as an interaction curve on the structural capacity curve.
- Select the proper design cross-section to handle the imposed moment, M_{max}, from those cross-sections that are at or above the intersection of the two curves.
- In the design for p_{max}, for the design example, if instead of M_{max} a limitation on p_{max} were necessary, a "structural capacity curve" giving the maximum soil pressure allowed for each pile width versus log S provided for that section could be developed.
- In the design for L, alternatively a design approach that incorporates this pile length, L, as a variable would require the development of a set of structural capacity curves for piles of different lengths in the third step of the design procedure. Additionally, structural capacity curves can be developed for pile lengths of 10 and 40 ft. The intersections of the structural capacity curves with the imposed M (or non-dimensionalized as $M_{max}/(PL)$) structural curves yield the design section that must be used for each particular pile length.
- Evaluate (i) allowable soil pressure, (ii) pier moment, and (iii) displacement values. The PFrame analysis program determines the pier moments, displacements, and soil pressures in a laterally loaded pier.

  - The ultimate soil pressure as determined by Broms for an assumed rigid body rotation is

Cohesionless soils:	$p_{max} = 3K_p \gamma y$
Cohesive soils:	$p_{max} = 5q_u$

The allowable net soil pressure, p_{max}, should be less than 1/3 to 1/2 of the ultimate unit bearing capacity of the soil (i.e., FS = 2 to 3).

For sands, an approximation of the allowable soil pressure, p_{allow}, is

$$p_{\text{allow}} = 3K_p \gamma L/\text{FS}$$

where K_p = passive Rankine soil pressure.
For clays, $p_{\text{allow}} = 5q_u/\text{FS}$.
The program computes these values using FS = 2.
An increased pier width will lower the soil pressure.
- The allowable moment, M_a, in the pier is only approximated by the program using an allowable concrete stress of $f_c'/3$ and the gross section moment of inertia

$$M_a = f_a S_{xx} = (f_c'/4)I_{\text{gross}}/(\text{dia}/2)$$

For a circular section the section modulus is $\pi \, \text{dia}^3/32$ based on the gross concrete section.
The proper way to determine the working and ultimate load capacities of a rectangular and circular pier section is to look at it like a beam–column and use the program **PMEIX-VBA** presented in Chapter 8.
An application showing this approach is presented in **Example 18.5a** for a pier of a bridge bent.
- Lateral displacements at the top of the pier are usually limited to being less than 1 in depending on the structure that the pier supports.

18.10 Solution Exclusively in Excel Worksheet without VBA

Setting up the equations that need to be solved for a laterally loaded pile problem can be performed exclusively within the worksheet without using any associated macro like the VBA program **PFrame**. For a simple problem having a generalized configuration of beams and springs, such an approach would add the spring stiffnesses to the diagonal of the S matrix and develop deformations using the math functions MINVERSE available within the worksheet. Multiplying matrices using the math function MMULT yields the member forces along the pier.

When using the math Function MINVERSE from the Insert > Function > Math and Trig menu bar in Excel to invert a matrix, do not enter braces { }. Matrix algebra operations such as Transpose, Multiply, and Matrix Inversion cover arrays defined by a range of rows and columns. This function is entered by pressing Ctrl+Shift+Enter simultaneously. The braces will automatically be entered during the ShiftCtrlEnter step. They indicate an array formula.

This solution cannot be applied generally since the development of the SS matrix requires a simple spring–beam configuration. A single pier is just such a problem. **Example 18.7**, a concrete pier similar to **Example 18.5** but with the pier height above the ground surface, H, equal to zero, is solved using both **PFrame** (a) and this section's approach (b). Worksheet **Example 18.7b LLP-Excel-Gen** can handle the three loading and support conditions that exist in a general problem.

18.11 Point of Fixity

If the portion of a bent above the foundation level is to be analyzed rather than including the foundation and soil properties below grade, it is necessary to determine where the support should be applied to the column of the bent. The results of the analysis of a single-pier foundation are to be interpreted so support can be applied to the superstructure. The point of fixity of the pier foundation system needs to be determined from the single-pier analysis. This point of fixity can be defined as either:

- the point of maximum moment, M_{max}, in the pier where the pier displacement is equal to zero; or
- the point of inflection where the moment is equal to zero (second point if loaded into negative moment range) (see moment and displacement plots of any of the examples).

A true point of fixity would have no lateral displacement as well as rotation. It is best to use the point of maximum moment as the point of fixity because the point where the moment is zero is too conservative. Usually the first definition (maximum moment) is employed and the distance from the ground surface to this level is given as the point of fixity. **Example 6.6** shows how the pier–bent system analyzed in **Example 18.6** can be made into a frame-only system by using a point of fixity below the foundation level.

18.12 Pile Groups

For horizontal pile loads applied perpendicular to a line of piles, if such piles are spaced 2.5 diameters or more, no reduction in their lateral capacity occurs. For piles spaced from 2.5 to 8 diameters, a linear lateral reduction factor varying from 0.7 to 1 as the spacing varies from 2.5 to 8 diameters is applicable. This reduction can be applied by reducing the k_h value used to a factor αk_h, where α is the reduction factor.

18.13 Conclusions

The soil from the ground surface to depths of 30% of the pile height exerts a controlling influence on the engineering behavior of a laterally loaded pile. The stiffness and the effect of seasonal conditions on the soil in this zone will greatly influence the behavior of a laterally loaded pile. Therefore, with respect to reducing the deflection of the pier at the surface and the maximum moment in the pier, there is little benefit from having a stiff surface layer that extends deeper than 30% of the pile length.

From the moment diagram it is evident that steel reinforcement in a pier is predominantly required in the top third of the pier length and that its full area need not be extended to the bottom of the pier. Also it is not good practice to dowel or splice rebar or make a cold joint in the concrete pour where the moment is maximum, if possible.

18.14 Significant Aspects of Excel Worksheet and VBA Macro

The following features of an Excel worksheet are shown in the examples and VBA program associated with this chapter:

- Matrix Multiplication MMult
- Matrix Inversion
- Nested If Statement
- Sending Output data from VBA macro to worksheet and Plotting (charting) it in worksheet
- Double versus Single precision and the "Long" integer.

The worksheet cell output data consisting of pile moments, displacements, and soil pressure patterns adjacent to the pile is charted using that feature of Excel. The location of the output data must remain the same on all runs as the source for charting this data must begin at the same location associated with the spreadsheet cell data. Since the input data is developed automatically for a set pile model size, maintaining this same cell location on output is generally no problem.

 ## 18.15 Examples

18.15.1 Example 18.1 ADOT Lateral Load Test

A lateral load test was conducted by the Arizona Department of Transportation (HNTB, 1984) on a steel H-pile whose properties are presented below. By inputting various values of k_1, the value that produces the measured deflection under the applied test loads can be determined. In the test, the coefficient of subgrade reaction value yielding the measured top deflection is $k_1 = 120$ pci or 207 kcf.

Pile Properties: Geometry: length $= 45$ ft, section HP 14×117, $B = 14.885$ in ($B = 1$ used),
$A = 34.4$ in^2, $I = 1220$ in^4, $SM = 172$ in^3
Material: steel, $E = 30 \times 10^6$ psi $= 4.32 \times 10^6$ ksf, $f_y = 60$ ksi, $(f_b)_a = 24$ ksi, $M_a = 344$ k-ft.

Soil Configuration and Properties: A 40 ft depth of overburden consisting of clayey sand and sandy clay.
Tip extends 5 ft into a sand–gravel–cobble (SGC) layer $k_h = k_1 y$.

Top Support: Free.

Lateral Load: Design $= 40\,k$, maximum applied during test $= 80\,k$.

Results: Essentially, a linear lateral deflection response resulted from applying the lateral load at the top of the pile:

Top deflection (in)	Applied lateral load (k)
0.3	40
0.6	80

18.15.2 Example 18.2 Concrete Pier

An analysis of a short, small-diameter concrete pier is performed in this example problem. Data for the pile and soil is presented as follows:

Pier:

Length above grade $= 6$ ft
Length below grade $= 18$ ft
Reinforced concrete circular:

Dia $= 2$ ft	$A = 3.14$ ft^2	$I = 0.785$ ft^4
$f'_c = 3$ ksi	$M_{allow} = 339$ k-ft	

Soil: Sand $\phi = 30°$, $n = 1$.

Top Support: Free.

Lateral Load: Applied at top of pier, $P = 6\,k$, $M = 3\,k\text{-ft}$.

The program output, including a chart (plot) of the moment, displacement, and soil pressure results, is presented in the worksheet for this example. It can be seen from a comparison of the computed moments and soil pressures to the allowable values that they are acceptable.

This example shows that the soil from the ground surface to depths of 30% of the pile height exerts a controlling influence on the engineering behavior of a laterally loaded pile. This example also shows the increased soil pressures on the passive side near the top due to the effect of large displacements near the top of the pile. The calculated moment in the pier is essentially the same throughout the pier, regardless of the bottom support conditions.

18.15.3 Example 18.3 Timber Pile

An analysis of a 20 ft long by 12 in dia ($B = 6$–14 in) timber ($E = 1500$ ksi, $\sigma_a = 2$ ksi) pile subject to a 4 k lateral load at its free top end embedded in a sand ($k_h = 15/B$, $n = 1$) is shown in this design example from Sogge (1981). The structural curve and the interaction curve are created to find the optimum section.

Timber piles typically taper from the top down to their tip. If this taper is known for a specific timber pile then the moment of inertia for the pile can be varied along the pile by specifying many beam elements having different material properties.

18.15.4 Example 18.4 Concrete Pier 2 Soils

This example analyzes a pier that penetrates through a 12 ft thick sand layer 18 ft into a clay layer.

18.15.5 Example Concrete Pier of Bent done by Three Approaches

Example 18.5a Concrete Pier of Bent with Single Spring

This example takes one pier out of the concrete frame bent analyzed in **Example 18.6**:

Pier: Reinforced concrete $f'_c = 4$ ksi.
Length below scoured grade level = 60 ft.

Circular:	Dia = 4 ft	$A_g = 12.6$ ft^2	$I_g = 12.6$ ft^4	$M_{allow} = 1000$ k-ft
Soil:	Sand	$\phi = 30°$	$k_1 = 60$ kcf	$n = 1$

Load: Lateral, due to stream flow = 12.5 k.
Top Support: fixed against rotation
Results: The maximum service (working) stress moment capacity of a 48 in diameter pier under an 800 k axial load derived using PMEIX for the circular section is 1000 k-ft when stress in the steel reinforcing reaches 24 ksi. This capacity is considerably greater than the moment resulting in the piers of approximately 170 k-ft. The quantity of steel reinforcement can be reduced. This result will be refined by the model used in **Example 18.6**.

From the results of this pier analysis, the point of fixity, if taken as the point of maximum moment (100 k-ft), is at 8 ft depth below grade.

Example 18.5b DoubleSpring

Same as **Example 18.5a** with a spring on each side of the pier and FEF added to the result.

Example 18.5c DoubleSpring Imposed Displacement

Same as **Example 18.5a** with a spring on each side of the pier and initial overburden pressure applied by imposed support displacements.

18.15.6 Example 18.6 Bridge Bent with Piers

This example is a bridge bent comprised of four piers as in **Example 18.5**. The automatic input data development of the worksheet cannot be done for general lateral load systems with many piers. For a bridge bent comprising pier foundation systems and a superstructure of columns (upward continuation of pier foundations above grade) supporting beams, a general application of PFrame that models the piers and their lateral soil support by springs and the columns and beams, all in one model, is best. There is no need to try to separate out the piers, find their "point of fixity," and then use that knowledge to develop and analyze a separate structural frame model consisting of only beams and columns. Example 6.6 shows this approach whereby the bottom of the columns, the top of the piers, were fixed at their point of fixity. When personal computers were in their infancy with limited memory, such an approach was necessary to make the problem tractable. A better model, possible today, would be one that included the piers and their soil support as is created in this example.

Structure: Analyze in direction parallel to bent resisting streamflow.
                   Reinforced concrete $f'_c = 4$ ksi.
Pier: Spacing 20 ft on center.
Length above scour level $= 40$ ft
Length below scoured grade level $= 60$ ft.
Circular: Dia $= 4$ ft $A = 12.6$ ft^2 $I = 12.6$ ft^4 $M_{allow} = 1000$ k-ft
Beam: 4ft × 4 ft $A = 16$ ft^2 $I = 21.33$ ft^4
Soil: Sand: $\phi = 30°$ $k_1 = 100$ kcf $n = 1$
Hydraulics: Streamflow acting on piers – leading pier gets twice the force.
Pier scour depth $= 15$ ft.
Load: Vertical: due to DL + LL − 30 k/ft.
Lateral: Due to streamflow – 20 k on leading column, 10 ft below top beam,
10 k on other columns applied at same elevation.

   Change to VBA PFrame Program: Due to the large size of this model it is necessary to make some changes to the standard PFrame program. The integer NONB = NODOF × NBW for this model becomes $408 \times 105 = 42\,840$. This value exceeds the size of an integer, $-32\,768$ to $32\,767$. Therefore it is necessary to "DIM NONB as Long" as well as the integer variables associated with NONB, rather than as integers.

   A comparison of the results from **Example** 18.6 to Example 6.6 analyzed in Chapter 6, in which a bent consisting of columns and beams, supported by fixed supports at the bottom of the columns (top of the piers) at their point of fixity, to results from a model that includes the piers and their soil support below the bent, follows:

Bent	Bent and soil
$\delta_{lat} = 0.008$ ft	0.006 ft

The bent and soil has less displacement than the bent alone since the soil support is higher, making the length of the column shorter.

It is evident that changes in moment and displacement responses can vary by up to 8% due to a more flexible foundation resulting from using cracked-section moments of inertia. Decreases of up to 16% can be had due to lower resisting loads resulting on more flexible structures.

18.15.7 Example 18.7 Concrete Pier

Examples 18.7a and **18.7b** analyze the structure presented in **Example 18.5** with a height of the pier above the grade level modified to essentially zero (H = 0). The analysis is then performed in two ways: (a) employing the PFrame macro solution and (b) employing a solution that is totally contained within Excel and uses the math functions MMULT and MINVERSE to obtain the same result.

 Related Workbooks on DVD

Example 18.1 ADOT LatLd Test
Example 18.2 Concrete Pier
Example 18.3 Timber Pile
Example 18.4 Concrete Pier 2 soils
Example 18.5a Bridge Bent-Single Pier
Example 18.5b Double Spring
Example 18.5c Double Spring-Imposed Disp
Example 18.6 BridgeBentwithPiers
Example 18.7a Concrete Pier (Example 18.5 with H = 0) – Pframe Macro Solution
Example 18.7b LLP-Excel-Gen – Same as Example 18.7a – Excel Workbook

References

AASHTO (2010) *LRFD Bridge Design Specifications*, 5th edn, American Association of State Highway Transportation Officials.

ACI (1989) *Building Code Requirements for Reinforced Concrete*, ACI 318-89, American Concrete Institute.

Bowles, J.E. (1977) *Foundation Analysis and Design*, 2nd edn, McGraw-Hill.

Broms, B.B. (1964a) Lateral resistance of piles in cohesive soils. *Journal of the Soil Mechanics and Foundation Division, ASCE*, **90** (SM2), 27–63.

Broms, B.B. (1964b) Lateral resistance of piles in cohesionless soils. *Journal of the Soil Mechanics and Foundation Division, ASCE*, **90** (SM3), 123–156.

Clough, G.W. and Tsui, Y. (1977) Static analysis of earth retaining structures, in *Numerical Methods in Geotechnical Engineering* (eds. C.S. Desai and J.T. Christian), McGraw-Hill, pp. 506–527.

Cox, T.S. (1992) Program CurveFit, QBasic ver 2.25a (5/9/92), Easley, SC (based on Kolb, W.M. (1984) *Curve Fitting for Programmable Calculators*, 3rd edn, SYNTEC, Inc., Bowie, MD, Spiral bound, January).

Desai, C.S. and Kuppusamy, T. (1978) Procedure for a soil-structure interaction problem: analysis and evaluation. Computing in Civil Engineering, ASCE, Atlanta, GA, June, pp. 200–218.

Ghosh, S.K., Fanella, D.A., Rabbat, B.G. (eds.) (1996) *Notes on ACI 318-95, Building Code Requirements for Structural Concrete with Design Applications*, Portland Cement Association.

Habibaghi, K. and Langer, J.A. (1983) Horizontal subgrade modulus of granular soils, *Laterally Loaded Deep Foundations, Analysis and Performance*, STP 835, June, ASTM, pp. 21–34.

Halliburton, T.A. (1968) Numerical analysis of flexible retaining structures. *Journal of the Soil Mechanics and Foundation Division, ASCE*, **94** (SM6), 1233–1251 (Proc Paper 6221).

Hetenyi, M. (1946) *Beams on Elastic Foundation*, University of Michigan Press, Ann Arbor, MI.

Howard, Needles, Tammen & Bergendoff (HNTB) (1984) Advanced Test Pile Program, I-10, ADOT, June.

Poulos, H.G. and Davis, E.H. (1980) *Pile Foundation Analysis and Design*, John Wiley & Sons, Inc., Sections 7.2.1 and 7.2.2, 144–152.

Reese, L.C. (1984) *Handbook on Design of Piles and Drilled Shafts under Lateral Load*, FHWA-IP-84-11, FHWA, July. Program PMEIX Instructions and Listing, pp. 255–292.

Reese, L.C. and Allen, J.D. (1977) *Drilled Shaft Manual*, Structural Analysis and Design for Lateral Loading, Vol. II, FHWA, IP 77-21, July. Program PMEIX Instructions and Listing, pp. 189–224.

Sogge, R.L. (1981) Laterally loaded pile design. *Journal of the Geotechnical Engineering Division, ASCE*, **107** (9), 1179–1199.

Sogge, R.L. (1984) *Microcomputer Analysis of Laterally Loaded Piles*, STP 835, ASTM, pp. 35–48.

Sogge, R.L. (1992) Program LLP – Finite Element Analysis of Laterally Loaded Reinforced Concrete Piers, distributed by the Concrete Reinforcing Steel Institute (CRSI).

Terzaghi, K. (1955) Evaluation of coefficients of subgrade reaction. *Geotechnique*, 297–326.

19

Cantilevered and Anchored Sheet Piles

19.1 Cantilevered Sheet Piles

Cantilevered sheet piles are solved by assuming that soil resistance only comes from the soil below the dredge level. A Coulomb active pressure is applied to the active soil side and a Coulomb passive resistance below the dredge level. The calculations for maximum moment and depth of embedment come from solving the equilibrium equations for these quantities.

19.2 Beam–Bar Finite Element Model for Cantilevered Piles

The beam–bar model similar to those employed for laterally loaded piles is used. The only difference is that where the soil above the dredge level exists, no springs are used. Instead a Coulomb active pressure state is used for this soil and the corresponding joint loads are applied to the beam section above the dredge level.

On the backfill side above the dredge level, a loading from the soil created by an active soil pressure distribution is assumed. This pressure assumption is not too far from the truth as no arching would occur without a tie rod creating a top bridge point for the soil. Even so, the use of springs in a stress state shy of an at-rest pressure distribution is difficult to implement. A Coulomb pressure distribution that incorporates wall friction should be used rather than the Rankine soil pressure that does not address the state of friction between the wall and the soil.

19.3 Anchored Sheet Piles

The usual procedure in solving for the depth of embedment and the tie-rod force is to use the free-earth support method that assumes the wall is rigid with passive pressure

Solutions for Soil and Structural Systems using Excel and VBA Programs, First Edition. Robert L. Sogge.
© 2012 John Wiley & Sons, Ltd. Published 2012 by John Wiley & Sons, Ltd.

below the dredge level on the resisting side and active pressure on the backfill side. Equilibrium equations are set up and solved for the embedment depth.

In his many tests on models, Rowe published a method for designing anchored sheet piles in which factors are applied to the results determined from a free-earth support analysis. Tschebotarioff also developed methods for design based on his tests on large-scale models. These factors account for the flexibility of the pile. The degree to which such results are reduced is dependent in essence on the soil–structure interaction ratio of the soil to the structural stiffness.

19.4 Beam – Bar Finite Element Model for Anchored Sheet Piles

A beam–bar finite element model for the analysis of an anchored sheet pile system is identical to the model developed for a cantilevered sheet pile with the addition of a tie rod to anchor the top of the pile. For an assumed embedment configuration, the beam–bar model consists only of the region below the dredge level as all support to the pile comes from this embedded portion. An active pressure distribution above the dredge level on the backfill side of the sheet pile is assumed. Corresponding joint forces are applied to the beam model. As with cantilevered piles, a Coulomb pressure distribution should be used. The tie-rod force can be simulated either with a force applied to the tie-rod level joint or, with the addition of a spring at the tie-rod level, having a stiffness of EA/L.

The approach of applying the loads from an active pressure distribution above the dredge line rather than using springs to represent the soil in this region is done for two reasons. The stress composition in the soil experiencing active movement that includes arching between the increased soil pressure intensity at the tie-rod level and below the dredge level is very complex. Also the stress conditions that are less than the at-rest in-situ pressure state are difficult to model in this region using a linear spring. For these reasons they are removed and a single spring is used only to represent the tie-rod portion of the structure.

The acceptability of the calculated depth can be determined from the soil pressures resulting below the dredge level in a manner similar to that detailed for laterally loaded piles in Chapter 18. To ensure that the design is safe, it is recommended that various configurations with ranges on soil parameters, ϕ, c, γ, and surcharge loads, q, be analyzed.

19.5 Soil Strength Representation

A coefficient of horizontal subgrade reaction is used to supply the stiffness to the springs below the dredge level. The following equation can expressly be used:

$$k_s = a + bz^n$$

On the active side the depth z can be measured from the ground surface level and not from the dredge level.

19.6 Examples

19.6.1 Example 19.1 Cantilevered Sheetpile

This example shows how the worksheet for a general laterally loaded pile can be developed into a model for a cantilevered pile. Since this cantilevered pile has no tie rod attached to it, the area of the tie rod is set at a very small number. The pile is 25 ft in length with 11 ft embedded below the dredge level, which is the same as the low-water level. The computed 24 k-ft/ft moment occurs below the dredge level.

19.6.2 Example 19.2 Anchored Sheetpile with Tie Rod

This example is for a 41 ft long sheet pile embedded 10 ft below dredge level. A tie rod is attached to the sheet pile to anchor its top section. The worksheet is set up to apply a spring stiffness equivalent to the tie-rod stiffness. Alternatively, a force of so many kips could be applied to the 1 ft section of sheet pile. The deflection for that load application can be checked and the load can be changed to bring the deflection within acceptance levels.

19.6.3 Example 19.3 Anchored Sheetpile-Rowe Calcs

This example is essentially the same as **Example 19.2** that is solved by "longhand" methods using the charts developed by Rowe. Its use is more as a comparison to the output of **Example 19.2**.

 Related Workbooks on DVD

Example 19.1 Cantilevered Sheetpile
Example 19.2 Anchored Sheetpile with TieRod
Example 19.3 Anchored Sheetpile-RoweCalcs

Further Reading

ASCE (1996) *Design of Sheet Pile Walls*, adapted from US Army Corps of Engineers Technical Engineering and Design Guide No. 15, American Society of Civil Engineers.

Bjerrum, L., Clausen, C.J.F., and Duncan, J.M. (1972) *Earth Pressures on Flexible Structures*, Publication No. 91, Norwegian Geotechnical Institute, pp. 1–28. (Also in Proceedings of the 5th European Conference on Soil Mechanics and Foundation Engineering, Madrid, 1972).

Bowles, J.S. (1977) *Foundation Analysis and Design*, 2nd edn, McGraw-Hill.

Halliburton, T.A. (1968) Numerical analysis of flexible retaining structures. *Journal of the Soil Mechanics and Foundations Division, ASCE*, **94** (SM6), 1233–1251 (Proc Paper 6221).

Rauhut, J.B. (1966) A finite-element method for analysis of anchored bulkheads and anchor walls. PhD Dissertation, University of Texas, Austin.

Rowe, P.W. (1952) Anchored sheet pile walls. *Proceedings of the Institution of Civil Engineers*, **1** (Part I), Paper No. 5788, 27–70.

Rowe, P.W. (1955a) A theoretical and experimental analysis of sheet-pile walls. *Proceedings of the Institution of Civil Engineers*, **4** (Part I), Paper No. 5989, 32–69.

Rowe, P.W. (1955b) Sheet pile walls encastre at anchorage. *Proceedings of the Institution of Civil Engineers*, **4** (Part I), Paper No. 5990, 70–87.

Rowe, P.W. (1955c) The flexibility characteristics of sheet pile walls. *The Structural Engineer*, **34**, 150–158.

Rowe, P.W. (1957) Sheet-pile walls in clay. *Proceedings of the Institution of Civil Engineers*, **1** (Part 1), Paper No. 6201, 629–651.

Sogge, R.L. (1974) Finite element analysis of anchored bulkhead behavior. PhD Dissertation, University of Arizona, Tucson.

Tschebotarioff, G.P. (1949) Final Report on Large Scale Earth Pressure Tests with Model Flexible Bulkheads, Princeton University.

Tschebotarioff, G.P. (1964) Design and construction of flexible retaining structures. Presentation, Chicago Soil Mechanics Lecture Series.

Tschebotarioff, G.P. (1973) *Foundations, Retaining and Earth Structures: The Art of Design and Construction and its Scientific Basis in Soil Mechanics*, 2nd edn, McGraw-Hill.

Tschebotarioff, G.P. and Ward, E.R. (1958) Measurements with Wiegmann inclinometer on five sheet pile bulkheads. Proceedings, Fourth International Conference on Soil Mechanics and Foundation Engineering, London, vol. 2, pp. 248–255.

USS (1972) *USS Steel Sheet Piling Design Manual*, United States Steel, April.

20

Buried Arch Culverts (Tunnels)

The analysis of buried arches and tunnels creates solutions that are somewhat similar to those for buried laterally loaded piles and piers. The structure is different by being more related to a frame than a linear beam, thus it will have a different stiffness. An arch can described as being flexible as a corrugated metal pipe (CMP) or multi-plate structure or rigid in a concrete culvert (Krizek *et al.*, 1971; Kay and Abel, 1976; Krizek, 1977; Daemen and Fairhurst, 1978; Bulson, 1985; Moser, 1990; Abdel-Sayed *et al.*, 1994; McGrath and Howard, 1998; McGrath *et al.*, 2002).

20.1 Theory: Classical Elasticity Formulation – Burns and Richard Solution

The differential equation solution that governs the behavior of beam-on-elastic foundations (Chapter 16) and laterally loaded pile structures (Chapter 18; Hetenyi, 1946) does not apply directly to a buried elastic circular cylindrical shell embedded in an isotropic elastic medium loaded by a uniform surface pressure. One of the last classical elasticity solutions to be developed was by Burns and Richard (1964) for just such a system. The equation solution presents the stresses and displacements in both media. These solutions are what are referred to as classical solutions since they are closed-form solutions to the formulation of equilibrium, force–deformation, and compatibility equations representing the elastic media.

For this solution the load is uniformly distributed over the infinite soil surface extent above the inclusion buried deeply in the infinite medium. The solution is applicable to deeply buried conduits. Practically this depth is 1–2 diameters. The stress conditions due to the soil weight prior to the application of any surface load must be added to those stresses obtained from this analysis for a surface loading. Thus the burial depth does not enter into the solution except when computing the overburden pressure. Burns and

Solutions for Soil and Structural Systems using Excel and VBA Programs, First Edition. Robert L. Sogge.
© 2012 John Wiley & Sons, Ltd. Published 2012 by John Wiley & Sons, Ltd.

Richard (1964) presented solutions for the no-slippage and full-slippage cases, for the stress condition of zero shear stress at the interface.

The workbook **Burns&RichardEqns** provides a solution to the Burns and Richard equations for thrust, moment, and displacement in the shell and stresses and displacements in the soil around a buried cylindrical structure of given configuration and surface loading p. The interaction that is inherent between the cylinder structure and the surrounding soil is developed from the results of the equations.

20.2 Soil–Structure Interaction

The non-dimensional shell–medium interaction parameters consist of the circumferential extensional flexibility ratio related to

$$\{M_s \, (1 + K)\}/(EA/R)$$

and the circumferential bending flexibility ratio related to

$$\{M_s \, (1 - K)\}/(6EI/R^3)$$

M_s is the constrained modulus of the soil defined from the elastic constants of the soil, E_s and v_s, as

$$M_s = E_s(1 - v_s)/\{(1 + v_s)(1 - 2\,v_s)\}$$

and K, the lateral stress ratio,

$$K = v_s/\{(1 - v_s)\}$$

The bending stiffness of a shell per unit length, EI, applicable for plane-strain conditions, is $\{E/(1 - v^2)\}(t^3/12)$. The extensional stiffness per unit length, EA, applicable for plane-strain conditions of a shell, is $\{E/(1 - v^2))\}t$, where v is Poisson's ratio and t is the thickness, both for the shell structure.

The bending stiffness of the circular shell is related to EI/D^3. So, again, as was evident in Chapters 16 and 18, the behavior of such buried structures is related to a soil stiffness ratio, in this case being

$$M_s/(EI/D^3)$$

In the case of a buried culvert the equation is the dimensionless quantity

$$E_s/(EI/\text{span}^3)$$

(Allgood, 1972, p. 59). The units for I are ft^4/ft and those for E_s are k/ft^2. Note that the soil elastic modulus, E_s, and not the coefficient of subgrade reaction, is specified.

Using the constrained or confined compression modulus as defined above, Allgood (1972) set boundaries on the soil–structure stiffness ratios that define the performance of a buried cylinder of diameter D and stiffness parameters EI as

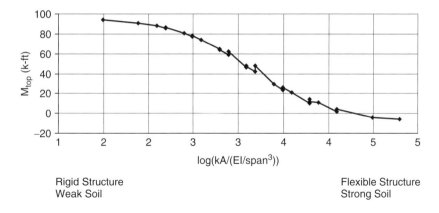

Figure 20.1 SSI from arch with lateral spring support (Example 6.8).

Flexible	$10^4 < M_s/(EI/D^3)$
Intermediate	$10^1 \leq M_s/(EI/D^3) \leq 10^4$
Rigid	$M_s/(EI/D^3) < 10^1$

Flexible cylinders are those for which bending flexibility presides over extensional flexibility, that is, the structure's mode of behavior is bending more than compressing extensionally.

A system consisting of a circular arch structure supported by a lateral spring representing the soil was presented in **Example 6.8**. The soil–structure interaction(SSI) curve presenting the results of the moment in the structure for various values of the spring constant is presented in Figure 20.1.

20.3 Beam–Bar Finite Element Frame Model

This distinction between flexible and rigid structures is important in selecting the applicable analysis approach. Flexible structures can be designed using both allowable stress design (ASD) and load and resistance factor design (LRFD) procedures that simulate and provide the results obtained from a full SSI analysis. Flexible culverts like CMP may be designed using different procedures that consider them as a compression rings (White and Layer, 1960).

Rigid culverts still represent a SSI problem, but due to the rigid behavior of the component material, usually reinforced concrete, it is necessary to limit the strength and thus the deformation characteristics of the structure so that they are compatible with the cracking properties of the concrete structure. By providing sufficient reinforcement to a concrete section, the cracking can be kept within limits. The deformation that occurs, though small, still results in the surrounding soil providing some resistance to the structural movements.

If concrete sections having large thicknesses and high reinforcement ratios are selected, an analysis of an elastic "frame" structure subjected to an assumed soil pressure can be performed. The use of the frame model will provide little difference from the results provided by a more complex SSI analysis of a similar rigid structure. The haunch and side support resistance provided by the soil may considerably exceed the at-rest soil pressures by a factor of 2–3.

An arched culvert structure can be analyzed using a beam–bar model and any general frame finite element analysis program. The response of a system composed of a culvert structure and the surrounding soil media to traffic loads is determined using the frame and spring element models of the soil–structure system. The system interacts as a unit in resisting the applied traffic and self-weight loads.

20.3.1 Frame with Spring Arch Culvert Model

A two-dimensional plane section perpendicular to the longitudinal axis of the arch structure is used. The span analyzed is a 1 ft wide transverse section, in the arch-spanning direction. Symmetry of arch geometry and loading is assumed as only half of the structure is modeled. A beam–bar analysis of the arch culvert with sidewalls and a slab foundation or a continuous footing with a cutoff wall is developed. Beam elements represent the arch barrel, supporting walls, foundation footings, and slab. Bar elements (springs) represent the stiffness of the spandrel fill, backfill, and native supporting soil. This soil resistance model of a buried structure is similar to that for a laterally loaded pile.

It may be desired to remove a spring from a model to apply a lateral load directly to the structure rather than through the spring. Rather than redo the entire model, a spring's effect can be removed from it by setting the stiffness equal to a very small number (10^{-10}). A zero value separates the support from the structure, in effect leaving it and its support hanging adjacent to the structure, and should not be used.

20.3.2 Coefficient of Lateral Subgrade Reaction

For shallow buried culverts the height of the soil above the crown cannot provide more vertical resistance than γ_{soil} times the soil cover height. Thus the variation in vertical pressure over the crown does not vary greatly from the overburden pressure. In modeling the structure, no spring to represent a soil reaction is used in the vertical direction above the joints of the arch.

The surrounding soil provides a lateral supporting stiffness to the structure. The true benefit of the soil to the structure enters in the lateral or horizontal pressure acting against the structure. In this direction a sizeable stiffness and pressure of two to three times the overburden pressure can be developed. The spring representing the lateral soil resistance is horizontal and not perpendicular to the arch surface.

The soil resistance adjacent to a buried arch structure can be represented using the same constant and coefficient of horizontal (or lateral) subgrade reaction that was employed for laterally loaded piles. Relations for k_h were developed using three approaches in Chapter 10 (Terzaghi, 1955). Soil resistance is represented by a coefficient of horizontal subgrade reaction for a free-draining backfill material having a linear strength representation that is depth dependent. Since, as noted in Chapter 10, $k_h = k_1 z$ is approximately equal to E, the expression k_{max} or kz can be substituted for the M_s or E_s parameter in the soil–structure stiffness ratio. The depth below the surface, z, can be the sum of the fill height above the arch crown and the height of the arch.

20.4 Vertical Loads

Vertical loads on the culvert consist of the following:

Design Truck Vehicle LL (HS20 with 32 k axle LL spaced 14 ft apart (AASHTO, 3.6.1.2.2)

Design Tandem Vehicle LL with 25 k axle LL spaced 4 ft apart (3.6.1.2.3)

With a dynamic load factor of $1 + 0.33(1.0 - \text{Fill Height}/8)$ (3.6.2.2)

Applied through tire contact area (one or two tires) of 20 in wide $\times$ 10 in long (3.6.1.2.5)

Lane loading 0.064 k/ft^2 over 10 ft wide lane

Fill material DL of 0.120 kcf.

The combination of DL from the fill height and LL from the traffic (HS20-44) which results in the same pressure on the culvert for various fill heights is presented in Figure 20.2 (from AISI, 1983, Figure 3-3, p. 100). This figure shows that the minimum loading on a culvert occurs when the fill over a culvert is approximately 5 ft in height above the structure.

20.4.1 Multiple Presence of LL

The multiple presence factor, m (3.6.1.1.2), could be taken as 1.0 depending on the distribution pressure overlap that develops. A distribution width in excess of the 10 ft

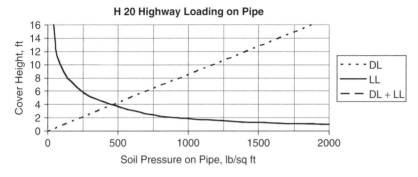

Figure 20.2 Soil pressure versus fill height for H 20 vehicle loading. (From American Iron and Steel Institute, 1983. Figure 3-3, p. 100).

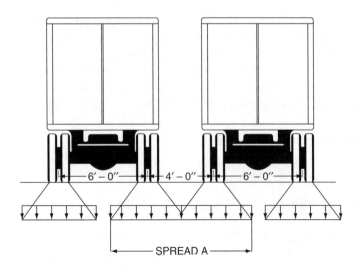

Figure 20.3   Overlapping adjacent lane traffic.

lane width occurs for shallow fills over structures having a span length larger than 16.67 ft. Adjacent lane pressure bulbs could then overlap requiring an m value of 1.2. Miller and Durham (2008) present such an argument. Figure 20.3 shows these stress overlaps as adjacent lanes can close their distance apart.

20.5 Distributing and Attenuating Vertical Live Loads

The concentrated or distributed axle LL can be attenuated by distribution through the earth fills in both the longitudinal and transverse directions of the culvert based on the center-to-center axle separation configurations (6 ft in the longitudinal and 4 ft in the transverse direction for a tandem axle) (AASHTO 3.6.1.2.3), wheel length (12 in) and width (20 in) (3.6.1.2.5), and soil area distribution factor (1.15 for granular, 1.0 otherwise) (3.6.1.2.6) as shown in Figure 20.4 (3.6.1.2.6).

20.5.1 Parallel to the Longitudinal Axis of Arch

The distribution of wheel loads in the direction parallel to the arch structure's longitudinal centerline for fill heights less than 2 ft is determined by distributing them as a line load over a distance of 8 + 0.12 Span (ft) (Article 4.6.2.10, AASHTO, 2010). If spans are greater than 16.67 ft, the distribution length is greater than a lane width and the stress bulbs will overlap.

For fill heights greater than 2 ft an equation similar to 2(20/12 + 1.15 Fill Height) could be used until the stress bulbs overlap at a Fill Height of 3.77 ft, and 6 + 20/12 + 1.15 Fill Height as indicated in Article 3.6.1.2.6 is applicable. It is probably better to use only

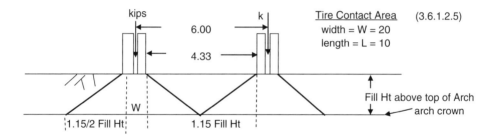

Figure 20.4   AASHTO attenuation of surface loads.

the latter equation for fill heights in excess of 2 ft to be consistent with the equation for distributing loads for fill heights less than 2 ft. Figure 20.4 shows these constructions.

20.5.2 In the Transverse Direction

In the transverse direction, similar equations can be developed for the design truck as

$$10/12 + 1.15 \text{ Fill Height}$$

and for the design tandem as

$$4 + 10/12 + 1.15 \text{ Fill Height}$$

In a finite element analysis (FEA) with general elasticity soil elements, these elements will automatically distribute applied surface load in the transverse direction.

20.5.3 Determination of p_v and p_h from Line Load

If the vertical surface load is a line load, it can be distributed in the transverse direction by using Boussinesq's equation. The attenuated vertical, p_v, and horizontal, p_h, soil pressure from an infinite vertical line load applied on the surface of semi-infinite mass (Poulos and Davis, 1974, p. 26) can be computed from

$$p_v = 2Py^3/\{p(x^2 + y^2)^2\}$$

(y = the vertical distance from a point below the ground surface to the centerline of arch section and x = the horizontal distance from a point below the ground surface to the centerline of arch section)

$$p_h = 2Px^2y/\{p(x^2 + y^2)^2\}$$

For p_v not to go infinite, the fill height must be more than 0.33 ft.

20.6 Horizontal K_o Pressure Load

The initial fixed-end force state on the structure that results in no initial structure movement is a $K_o\gamma z$ condition. Additional resistance that can be provided by the soil adjacent to the structure is related to the stiffness of the resisting soil, k_h, as quantified by the coefficient of horizontal (or lateral) subgrade reaction.

20.7 Load Application

In a frame type of analysis the DL of the soil, the lane loading, and the pressure from the concentrated axle loads, after attenuation, is applied directly to the structure as a concentrated load to each beam element joint or as a distributed load directly to each arch member. In a beam–bar frame analysis the beneficial horizontal component of the vertical LL is often ignored.

The horizontal component of the vertical soil DL from the at-rest state can be applied in two ways:

1. It can be imposed on the structure through the spring element representing the soil by applying a deflection to the spring element through its horizontal support that creates a K_o stress state. By this approach, the resultant horizontal loading or soil pressure distribution on the structure, which is dependent on the SSI, can be determined:

$$\text{Horz soil load} = K_o\gamma_s z$$

The horizontal springs are precompressed with an imposed deflection to provide the at-rest pressure, K_o, from the horizontal component of the vertical LL and soil DL:

$$\text{Deflection} = (K_o\gamma_s z) \times A/\{K_o A\}$$

applied as a joint displacement at each spring horizontal support.
2. Alternatively, the K_o at-rest state forces could be applied directly to the structure. It is tacitly assumed that no deflection occurs when the K_o at-rest force of $K_o\gamma z$ is applied to the joint. Then, as the structure moves, it would be resisted by the spring attached at the joint of the structure. The resultant stress state in the soil would be the at-rest K_o stress state plus the added or subtracted stress state computed for the spring.

A system consisting of an 18 ft span arch culvert supported by a slab foundation and loaded by an AASHTO highway loading was analyzed in two different ways: (i) by imposing the K_o displacement as a support displacement to the lateral spring, and (ii) by separately applying the K_o stress state force to the joint that the lateral spring is attached to. These approaches are shown in **Examples 20.1a** and **20.1b**.

An example of a 28 ft span arch culvert having the configuration of a cutoff wall rather than a slab foundation is shown in **Example 20.2 ArchCOW-28**. The VBA macro program is attached as a module to this workbook. **Example 20.3 ArchFtg-28** presents a model for a continuous footing foundation.

20.8 General Elasticity FEA Programs

A two-dimensional plane-strain FEA typically involves a finite element model of a transverse section perpendicular to the longitudinal axis of an arch structure that uses beam or plate elements to represent the arch barrel, supporting walls, and foundation slab and membrane, or two-dimensional solid elasticity elements to represent the spandrel fill and native supporting soil.

In models used by some researchers quadrilateral elements alone have been used to determine the behavior of the structure. These models tend to represent the structure in too stiff a manner as they are not as accurate as the beam elements.

This two-dimensional model ignores the distribution of wheel loads parallel to the longitudinal axis of the arch as it does not account for the continuity in the longitudinal direction, making this analysis very conservative. Differences between a two- and three-dimensional analysis model enter in only for traffic loads, and for such loads they arise only for culverts having shallow fills. A three-dimensional analysis is necessary to properly distribute such wheel loads.

If a linear soil model is used, a single-step application of live traffic load and dead load consisting of backfill soil around the arch body force of the concrete structure can be used. Linear representation of soil strength can be perfectly adequate as results show that loads on the soil are low and within the linear range of the soil.

Where interface elements are present, they should be used. A no-slip condition at the interface between the soil and the concrete arch is used if interface elements are not present. Esser, Sogge, and Richard (1974) have shown that for shallow-buried rigid conduits, the largest angle of friction developed at the interface is 10°. The Burns and Richard (1964) solution for buried tunnels showed that moments for the no-slip condition are 10% less than for the full-slip condition and that the maximum normal forces for the no-slip condition are 20% greater than for the full-slip condition.

A typical commonly used FEA program to analyze a buried arch culvert along with other shapes is program CANDE-2007-2011 (Mlynarski *et al.*, 2011) available on the Transportation Research Board (TRB)/National Cooperative Highway Research project (NCHRP) web site. The mesh extent and gradation to be used should be similar to those shown in the CANDE-2007 manual (pp. 5-123–5-127). If the structure and loading are symmetrical, half of the model can be used with appropriate boundary conditions. Many published solutions are available to verify an analysis. Similar models have been used by Krizek *et al.* (1971), Allgood (1972), Duncan (1979), and Katona and Vittes (1980) for buried arch structure systems for many research and practical problems.

The only reason such programs are not included here is because of the requirement of a mesh generating program for generating and displaying the results. An idea of what is involved is presented by the program **Flownet** for which a mesh has already been developed and output placed on it. A comparison of results using a beam–bar model

Table 20.1 Comparison of arch culvert analysis results (results of PFrame analysis of Arch Culvert 18 × 5.8313 in thick 50 k axle load $m = 1.2$).

R_1	R_2	M_{cL}	M_{haunch}	K_o
PFrame with springs – applied lateral displacement				
25	1	26	−27.7	1
25	2.75	22.5	−21.3	1
25	3	21.9	−20.34	1
20	2.9	21	−20.9	1
CANDE-2007				
25	3	21.8	−19.1	–
PFrame no springs – applied lateral force				
28	0	33.5	−35.4	0.5
28	0	32.5	−34.5	1
28	0	30.5	−32.8	2
28	0	28.4	−31.1	3
PFrame with springs – applied lateral displacement				
28	0	29.1	−32	0.5
28	0	28.7	−31.7	1
28	0	27.8	−31.1	2
28	0	27	−30.5	3

that uses springs for the soil to a full-scale elastic model as performed by CANDE-2007 is presented in Table 20.1.

20.9 SSI

In systems where soil exists as a supporting media of the load and the structure, the moments developed in the structure are dependent on the soil–structure stiffness ratio of $k_s/(EI/\text{Span}^3)$. An arch is such a system whose structural shape engages the soil's ability to partially carry load and support the structure. The stiffness of the soil is related to its elastic modulus or coefficient of subgrade reaction. The stiffness of the structure is dependent on the elastic modulus of the material, E, the moment of inertia of the cross-section, I, and its geometrical configuration represented by Span.

This dependency is shown in graphical form in Figure 20.5 for a 28 ft span arch culvert having different thicknesses. This figure shows the developed moment on the operating curve and the structural capacity on the capacity curve. The more flexible a structure is, the less the force that will be imposed on it, or the more load will be transferred to the soil material.

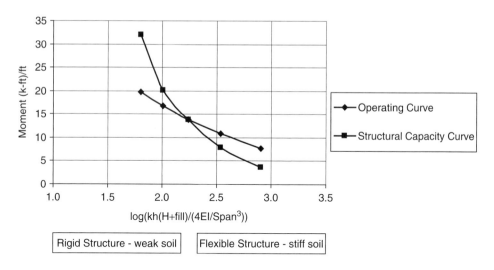

Figure 20.5   SSI curve for 28 ft span culvert.

20.10 Cracked-Section Considerations

With soil–structure systems where a transfer of loads from structural elements to soil elements occurs, and when designing for limit states, where the strains are much greater than where the steel first yields, in order not to be too conservative, the inclusion of cracked-section properties is appropriate for the structure at the design limit strength.

Prior to SSI analyses in which the soil component of a system was not represented or was not present, it never mattered about I_{cr} or I_g since their ratio, which governs the moments developed in the sections, would proportionally be approximately the same for all sections of the structure, resulting in little change in the load path.

This change in I as the section goes from an initial uncracked state to a final cracked-section limit state geometry should be reflected in an analysis of a SSI system by using an effective I representing the condition somewhere between those states. The following is a direct quote from Article 4.5.2.2, AASHTO LRFD (2010): "The stiffness properties of concrete and composite members shall be based upon cracked and/or uncracked sections consistent with the anticipated behavior." Such an analysis requires that those portions of the arch that are cracking be assessed a priori.

If a structural system is governed by stiffness for deflection control rather than strength control, the usual AASHTO span/800 limit criterion applies. It may be beneficial to use a reduced EI based on cracked-section properties as a check, even though the deflection criterion is applicable to service loads. Then even with the structure's increased flexibility a control on deflections is still maintained.

20.10.1 Cracked-Section Properties

For any material such as concrete, E is relatively constant throughout the load range. Since $M_{cap} = f_c' I/y$, the ultimate capacity is related to the stiffness of the section I and not the elastic modulus of the section. For the same-sized structure, the geometrical configuration is constant. Thus, in general, a structure's stiffness is solely related to its moment of inertia.

Usually when the moment capacity of a concrete element is computed, the equation $M_{cap} = f_c' I/y$ is not used. Instead, an equation based on the failure state of the compressive stress distribution on the face of the section is used. This equation, used to compute the moment capacity, accounts for the relocation of the neutral axis as cracking occurs and the resulting reduction in I. The section's reduction in I_{cr}/I_g with rotation (ϕ) has been quantified by Reese (1984) in his research and his development of the program PMEIX as was shown in Section 8.4 (Ghosh et al., 1996).

20.11 Examples

Workbooks for each of these examples develop arch culvert data for input to the VBA program **PFrame**.

20.11.1 Example 20.1(a) ArchSlab-LatSprDispLimit LL; (b) Arch-Slab-JtLd&SprngLimit LL

These examples show results from the analysis of an 18 ft span $\times$ 5.83 ft height arch culvert analyzed by (i) imposing the K_o displacement as a support displacement to the lateral spring and by (ii) separately applying the K_o stress state force to the joint to which the lateral spring is attached.

Both approaches give the same result.

Both of these examples have AASHTO 25 k tandem axle vehicle loads separated by 4 ft.

20.11.2 Example 20.2 ArchCOW-28

This example covers an arch culvert, 28 ft span $\times$ 7 ft height, supported on a continuous foundation with a cutoff wall.

20.11.3 Example 20.3 ArchFtg-28

This example covers an arch culvert, 28 ft span $\times$ 7 ft height, supported on a continuous foundation on both sides of a wall.

Related Workbooks on DVD

Burns&RichardEqn
Example 20.1a ArchSlab-latSprImpDispLimit LL
Example 20.1b Arch Slab-JtLd+SprngLimitLL
Example 20.2 ArchCOW-28
Example 20.3 ArchFtg-28

References

AASHTO LRFD (2010) *Bridge Design Specifications*, 5th edn, American Association of State Highway Transportation Officials.

Abdel-Sayed, G., Bakht, B., and Jaeger, L.G. (1994) *Soil-Steel Bridges*, McGraw-Hill.

AISI (1983) *Handbook of Steel Drainage and Highway Construction Products*, 3rd edn, American Iron & Steel Institute.

Allgood, J.R. (1972) Summary of Soil- Structure Interaction. Technical Report R 771, Naval CE Lab, Port Hueneme, CA, July.

Burns, J.Q. and Richard, R.M. (1964) Attenuation of stresses for buried cylinders. Proceedings of the Symposium on Soil-Structure Interaction, University of Arizona, September, pp. 378–392.

Daemen, J.J. and Fairhurst, C. (1978) Ground Support Interaction – fundamentals and design implications, Tunnels and Shafts in Rock, Army Corps of Engineers Report EM/10-2-290, Appendix A.

Duncan, J.M. (1979) Behavior and design of long-span metal culverts. *Journal of the Geotechnical Engineering Division, ASCE*, **105** (3), 399–418.

Esser, A.J., Sogge, R.L., and Richard, R.M. (1974) Stress distribution around shallow buried rigid pipes. *Journal of the Structural Division, ASCE*, **100** (1), 2349–2352.

Ghosh, S.K., Fanella, D.A., Rabbat, B.G. (eds.) (1996) *Notes on ACI 318-95, Building Code Requirements for Structural Concrete with Design Applications*, Portland Cement Association.

Hetenyi, M. (1946) *Beams on Elastic Foundations*, University of Michigan Press.

Katona, M.G. and Vittes, P.D. (1980) Soil-structure analysis and evaluation of buried box-culvert designs. *Transportation Research Record*, No. 878, 1–7.

Krizek, R.J., Parmalee, R.A., Kay, J.N., and Elnagger, H.A. (1971) Structural Analysis and Design of Pipe Culverts. Report 116, National Cooperative Highway Research Program (NCHRP)

McGrath, T.J., Moore, I.D., Selig, E.T., Webb, M.C., and Taleb, B. (2002) Recommended Specifications for Large-Span Culverts, NCHRP Report 473, Transportation Research Board – National Research Council, Washington, DC.

Miller, L.J. and Durham, S.A. (2008) Comparison of standard load and load and resistance factor bridge design specifications for buried concrete structures. *Transportation Research Record, Journal of the Transportation Research Board*, **2050**, 81–89.

Mlynarski, M., Katona, M.G., and McGrath, T.J. (2011) CANDE-2007 with 2011 Upgrade, Culvert Analysis and Design, User Manual and Guideline, Solution Methods and Formulations, Tutorial of Applications. Report 619, National Cooperative Highway Research Program (NCHRP). Downloadable from www.trb.org.

Poulos, H.G. and Davis, E.H. (1974) *Elastic Solutions for Soil and Rock Mechanics*, Wiley-VCH Verlag GmbH.

Reese, L.C. (1984) Handbook on Design of Piles and Drilled Shafts Under Lateral Loads. Report No. FHWA-IP-84-11, FHWA, July. (Used for program PMEIX.)

Terzaghi, K. (1955) Evaluation of coefficients of subgrade reaction. *Geotechnique*, **5** (4), 297–326.

White, H. and Layer, J.P. (1960) The corrugated metal conduit as a compression ring. *Proceedings of the Highway Research Board*, **39**, 389–397.

21

The Arch Form

21.1 History of Arches and Vaults

Some of the first arches in history were those of the Egyptians and the Mesopotamians in their constructions in about 2500–3000 BC (Van Beek, 1987). The Etruscans were the first to extend their use to what is now Italy and the Romans popularized them in the 1300s with the Pt Vechio Bridge in Florence. In the United States, the masonry Washington Aqueduct with a span of 220 ft and rise of 57 ft and the stone arch bridge that was constructed in 1883 to support a railroad crossing over the Mississippi River were impressive applications in construction using these materials.

The first concrete arch with steel reinforcing bars was the 20 ft span structure in Golden Gate Park by Ransome in 1889. Later, the Tunkhannock Viaduct was constructed in 1915 of 10 180 ft spans of reinforced concrete and the Washington Memorial arch bridge structure in 1932.

While earth-filled reinforced concrete arch culverts and bridges were popular from the turn of the last century until the 1930s, due to their being difficult to analyze and their requirement for curved surface forms, they fell out of use. Recent specialized curved metal forms and precast shapes along with computer analyses have increased their application.

21.2 Arch-Shaped Configurations

Various shaped arches are selected to carry the dead and live loads in an efficient manner and provide an aesthetically pleasing structure. The intent of selecting an arch-shaped configuration is to minimize the moment in the section compared to a straight section, minimize the thickness of the arch member, and develop the very helpful force, thrust.

The form of the compressive line of thrust through an arched section can take many paths described by a group of mathematical expressions. Usually the load path will

Solutions for Soil and Structural Systems using Excel and VBA Programs, First Edition. Robert L. Sogge.
© 2012 John Wiley & Sons, Ltd. Published 2012 by John Wiley & Sons, Ltd.

move in a shape minimizing the potential energy and assume some shape for which the expression used to characterize it is only an approximation. Where the centroid of the section coincides with the line of thrust, no moment will exist in the member.

21.2.1 Circular, Elliptical, Parabolic, and Catenary Shapes

The forms for the construction of arched culverts may be circular, elliptical, parabolic, catenary, sinusoidal, or a series of straight line arc segments. Such geometric shapes are described by the following algebraic equations:

Arc shape	Equation describing funicular shapes
Circular	$x^2 + y^2 = R^2$ R = radius of circle
Elliptical	Usually developed from a three-centered circular arc construction
Parabola	$y = -4h(x/\text{span})^2 + 4h(x/L)$ Uniform load along the horizontal suspension bridge
Catenary	$y = c\cosh(x/c) = c(e^{x/c} + e^{-x/c})/2$ Uniform load along curved arc length-like hanging rope
Hyperbolic cosine	$y = a\cosh[2x/L\cosh^{-1}\{(a+H)/a\}]$ Arc follows zero-moment curve for spandrel-filled structure
Straight lines	Formed from segments of a circular arc

Beer and Johnston (1962) present the parabolic and catenary equations as they are used to describe cables with various loadings. The workbook **GeomShapes** plots the various arch-shaped configurations discussed.

21.2.2 Inglis Equation for Spandrel-Filled Arch

Spandrel-filled arches can be designed so that the line of thrust due to the dead loads will lie close to the centerline of the arch and thereby produce small moments. C. E. Inglis in the early 1900s (Heyman, 1982, specifically pp. 46–49) derived the hyperbolic cosine equation given that defines the geometrical shape of a zero-moment curve for an arch loaded by the dead load of fill soil above its spandrel. The shape of this arch for the dead load line of thrust conforms to the shape of a modified catenary curve:

$$y = a\cosh[2x/L\cosh^{-1}\{(a+H)/a\}]$$

Inverse hyperbolic cosine $= \cosh^{-1}\{(a+H)/a\} = \ln[(a+H)/a + \{(a+H)/a\}^2 - 1\}^{0.5}]$

where
    a = height of spandrel fill above arch (ft) = distance from grade to centroid of arch (ft) (includes half of arch thickness at the crown)

H = arc height (ft) (includes half of arch thickness at crown)
L = arch span (ft)
x = horizontal distance from center of span (ft)
y = vertical distance from top of grade to zero-moment line (ft).

As long as the curve described by the Inglis equation is contained within the arch structure (or in the soil where arching in the soil supports the load) the arch will be stable and no failure can occur. Any moments that may develop due to the application of surface live loads must be considered separately.

An Excel workbook solution, **T-Circ, Ellipse, Cat&Par**, provides a solution to the Inglis equation for the moment and thrust in a structure that is developed due to soil in the spandrel zone area.

21.2.3 Segmented Arch Shapes

Construction of a curved shaped arch is difficult as specialized forms are required for its construction. A more manageable forming system would be to remove the arched shape from the bottom of the member where it is difficult to form and replace it with a chord of the circular arc. Also, an arch can be created by inclining flat straight slab sections into an arch shape. The addition of material beyond an arch shape between the chord of the arched stress path and the perimeter of the structure often does little to enhance the load capacity of the structure.

Figure 21.1 shows that where the circular arc line is closest to the member face, it is in the region where the moment is at a minimum (almost at the point of inflection where it is equal to zero). The circular arc section is well contained where the moment is greatest on the ends. An optimized member section would follow the circular arch with two or more parallel straight-lined segments containing the circular arc.

21.3 Force Determination for Various Shaped Arches

A very simple and quick solution approximation for the vertical and horizontal reactions and the thrust in a circular, elliptical (formed by a three-centered circular arc), catenary, and parabolic equation, and the Inglis equation, is included in the workbook **T-Circ, Ellipse, Cat&Par**.

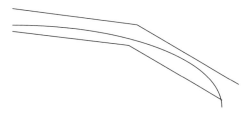

Figure 21.1   Optimized straight line section containing thrust line.

21.3.1 Parabolic Arch Solution Using Leontovich Equations

Leontovich (1959) published simple equation solutions for symmetrical two-hinged and fixed-support parabolic arches subject to loads of:

- Vertical complementary parabolic load over entire arch
- Vertical uniform load over entire arch
- Vertical concentrated load at crown.

These equations and their solutions are presented in the workbook **T&M Par LeontovichEqns**.

Alternative equations that include the effects of shear and axial deformation are presented. These solutions are applicable to parabolic-shaped arches having height to span ratios less than or equal to 0.2. The workbook **T&M Par LeontovichEqns** contains these equations.

21.4 Arch Engineering Considerations

Aside from being aesthetically pleasing, the benefit of an arched structure configuration to a structural member is that thrust (H) develops (see Figure 21.2). Two benefits of this thrust are as follows:

1. Thrust reduces the moments in the structure compared to moments in a flat structure as is evident by simple statics. The horizontal component of the reactive thrust significantly reduces the positive moments (bending stress) at the crown, and, similarly, the negative moments are reduced in the haunch section above the wall due to the traffic LL and soil DL applied to the structure.
2. By engaging the soil on the side, the developed thrust compresses the end section of each component member of an arched structure and forms an arched stress path within the member that further helps in resisting the applied load to reduce the moment in the section. Loads applied even to a linear or slab arced section will develop an arced load path shape within the structure as it naturally tries to minimize the bending and shear stresses (see Figure 21.3). The resulting compressive arc within the structure forms regardless of the geometric configuration of the external surfaces of the structural component.

Elliptical sections comprising three-centered circles are tangent to the wall at their point of intersection and do not develop the thrust pattern of flatter arcs. Shapes formed by an arc of a single circle, though easy to construct and to create different span forms from, have the disadvantage of their less steep inclination near the wall section that develops more thrust and thus more shear in the top portion of the vertical wall section.

As an arch culvert deflects vertically, thrust is developed in the arch structure as the passive resistance in the earth backfill counteracts the lateral deflection resulting

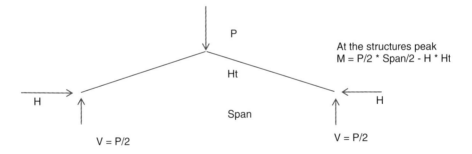

Ht

Span

At the structures peak
M = P/2 * Span/2 - H * Ht

H

H

V = P/2

V = P/2

Figure 21.2   Horizontal component of thrust developed in an arched section.

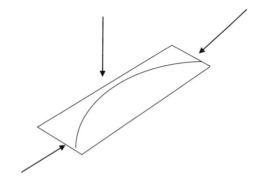

Figure 21.3   Compressive thrust line within the flat section.

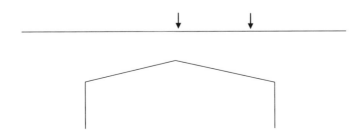

Figure 21.4   Attenuation of surface loads.

from the applied vertical loads. In a state of extreme overload the arch culvert cannot collapse without a lateral deformation of the block of soil behind the walls that is great enough to allow the arch to snap through. Even though three hinges may form in the culvert, the arch structure will still be a viable structure, supported in a non-collapse mechanism by the lateral pressure from the surrounding backfill. Until four hinges develop, the arch is a stable mechanism.

The structure provides its greatest resistance to loads applied directly over its peak. As loads deviate from the apex, more soil fill exists between the roadway surface where the load is applied and the structure. The more the fill, the greater the attenuation of the applied load. The moment also decreases as the load application point is distanced from the apex as shown in Figure 21.4.

21.5 Structural and Hydraulic Efficiency

The arch form, derived from the compressive load–resistance path, follows the function of minimizing the bending moment in the structure and embodies beauty in aesthetic arch configuration shape. A properly designed arched structure has a form that closely follows the load–resistance path and contains or envelops the resultant line of thrust of the applied loads, thus producing a moment of zero or nearly zero in the arch section. It is the most efficient configuration as it results in less bending and a greater compressive force. The predominant stress inside the arch is compression. The actual arched shape that the stress path assumes is irrelevant as long as it is confined in the section.

An arch structure configuration results in the most efficient use of concrete material since it carries loads in axial compression rather than by bending. Materials are much more efficient in carrying compressive stresses than bending stresses since the amount of reinforcing steel can be minimized. Even unreinforced concrete is capable of resisting large compressive stresses. Unlike a box culvert that is designed merely for bending capacity of the structural members, arched members are designed with the benefits of being beam–columns.

Another structural advantage of the arch shape is in the manner it resists loads without developing much shear stress. The arch shape and the horizontal end reactions greatly reduce any shear force that may develop in the section due to the vertical loading by resolving it into thrust within the arch. Shear values for a flat slab under similar loading would be significant. The soil restraint that develops the thrust in the curved top creates an arch action that increases the section's capacity to carry vertical loads.

The most efficient structural systems are those that develop thrust and then engage the surrounding soil to resist the thrust and thereby carry a portion of the load. These systems, known as soil–structure systems, are also the most difficult to analyze since they require a soil–structure interaction analysis. The results of that analysis are dependent on the soil–structure stiffness ratio to provide the portion of the load carried by the structure and the soil.

Not only are arch culverts structurally efficient, but one or two long-spanning arch structures can readily substitute for multi-barrel box culverts in an hydraulically efficient manner. This substitution permits a more efficient hydraulic flow by eliminating center walls that trap debris during flood events and by greatly reducing the hydraulic radius of the flow area.

21.6 Soil–Structure Interaction

If the thrust produced by the structural configuration is not resisted by the structure itself then another reactive force must resist it. In an arch culvert system, soil provides this resistance to the developed thrust and acts like a cable tensioned between the bottoms of the arch. The soil reaction in effect post-tensions the structure, significantly reducing the moments (bending stress) in the structure and increasing its design capacity.

An interaction between the soil and structural components develops within the system. Applied dead and live loads are resisted by the soil sharing the load with the structure. Another important aspect of soil–structure interaction is that the soil surrounding the structure minimizes load-induced deformations by providing confinement support of the structure.

Soil–structure interaction can be defined as follows. The culvert (structure) and the enveloping soil mass, working in tandem as a composite unit, are one of the most remarkably synergistic systems in engineering. When the soil is properly compacted around the culvert, the load-carrying capacity of the culvert–soil system far exceeds the individual capacity of either component by itself. Analysis of soil–structure interaction is the recognition that both the culvert and soil are structural components. Thus, the amount of load carried by the culvert is dependent on the soil stiffness relative to the culvert stiffness for various modes of deformation.

What does this definition mean exactly? It states that for a load applied to a structural system consisting of a structural and soil component, the load will be divided between the two components according to the stiffness of each. The principles of soil–structure interaction show that a longer span structure, which is more flexible, can develop relatively less moment than a short-span structure, which is relatively rigid, since less load is transferred to the surrounding soil by the stiffer structure. Therefore, relatively stiff short-span structures will not receive any reduction in moment from their flexibility and are going to behave in a more rigid manner.

As a structure and its surrounding soil interact to carry the imposed live traffic and dead earth loads, the buried structure and the adjacent soil will carry them in proportion to their stiffness. The stiffer component will carry a greater proportion of the total load than the more flexible component. For the same structure geometry, the more flexible structure will carry less of the imposed load than a more rigid structure since it engages the soil and distributes the loads into it. Any additional loading on a rigid structure that does not noticeably deflect will be carried by a continuously increasing load on the structure. A rigid structure, by sustaining more of the imposed load, can gain load up to the point where it finally cracks.

A plot showing this soil–structure interaction that results from an elastic finite element analysis (FEA) of arch culverts of various spans and thicknesses is presented in Figure 21.5. Using these FEA techniques, a thin, efficient, cost-competitive concrete arch section that works in conjunction with the soil to carry imposed loads is derived.

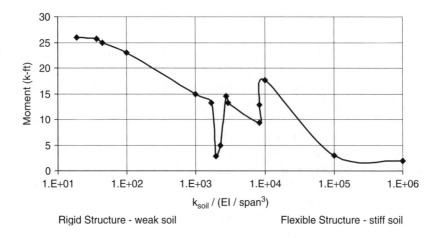

Rigid Structure - weak soil   Flexible Structure - stiff soil

Figure 21.5   Soil–structure interaction for arch culverts.

21.7 Flexible versus Rigid Structures

Flexible culverts are usually constructed of materials like aluminum or steel in a pipe or plate form or of plastic pipe. One common application is corrugated metal pipe or CMP. These materials can accommodate the large strains that result from the structure deforming in its interaction with the surrounding soil on which it is dependent to hold its shape. A rigid structure classification is typical of concrete pipes and culverts.

As to the use of rigid concrete or flexible metal in a culvert structure, it is beneficial to discuss the different properties of each material. Both materials can be used in similar applications. A flexible pipe is, by definition, a pipe which will deflect when subjected to external loads. A rigid structure does not rely as much on deformation to develop its carrying capacity.

This deflection that occurs in a metal structure under load can be advantageous as it demonstrates the load capacity and flexibility of this material. Flexibility in buried pipes is a desirable attribute as it engages the soil support. This interaction between the structure and how it relates to the neighboring soil, thereby establishing a functional soil–structure composite system, is key to its performance.

Under load, even the most flexible structures may only experience small amounts of deflection, if the soil backfill around them is properly installed. But it is in this state that the soil is engaged to support the structure and prevent further deformation. Both flexible and rigid-type structures require that the backfill be properly selected and compacted.

The more the structure deforms, the more the stresses in the surrounding soil adjust to accommodate this deformation. One deformation pattern that leads to a soil "arch" forming above the structure results from a downward vertical deformation of the structure of or settlement of the structure. In this case, the soil bridges or arches between the non-settling soil on each side of the arch structure to reduce the vertical load applied to the structure.

Flexible structures are less susceptible to differential settlement of their foundations and have the structural characteristic, unlike a rigid structure, of deforming without cracking. If overstressed, a flexible structure can get to the point of where the deformations are no longer sustainable and are excessive.

21.8 Failure Patterns and Deflections

A definition of failure is difficult to formulate since loads have to be excessive before failure is induced in a buried arch structure. The problem with using a very thin concrete section for a culvert is usually not shear or bending strength or buckling capacity but cracking which can result from excessive deflections of a concrete structure. A performance limit that defines the acceptable level of cracking of a concrete structure would be an appropriate design criterion. A criterion such as cracking which is deemed excessive by visual assessment will usually suffice.

Usually excessive cracking can be avoided by ensuring that the serviceability deflections developed under a 32 kip axle load times the dynamic load factor are less than span/800 (AASHTO, 2010, Article 2.5.2.6.2). In performing such a service load analysis just for live loads, the analysis should be performed without the soil DL applied but with supporting soil in place. By meeting the span/800 deflection criterion, crack widths should be limited to less than 0.01 in. Also, AASHTO Article 5.7.3.4, Control of Cracking by Distribution Reinforcement, and Article 5.6.3.6, Crack Control Reinforcement, provide a reinforcing design approach to control cracking.

Large deflections are not as critical for CMP structures that do not crack and can more readily handle any large deflections that may result while the soil becomes engaged.

It is interesting to note that confined bridge deck slabs exhibit arching behavior similar to what occurs in our culvert system in that confined deck slabs do not tend to fail in moment as expected, but rather tend to fail in a compressive (punching shear) mode. Wheel loads are resisted differently in a confined slab as compared to an unconfined slab that resists loads by beam action. Such arching action in bridge deck slabs was first reported by the Ontario Ministry of Transportation (Csagoly and Lybas, 1989; Bakht, 1997).

Besides checking the usual bending and shear stresses developed, punching shear should be checked under the applied wheel load at a distance of $d/2$ from the load, where d is the distance from the extreme compression fiber to the centroid of the tension steel. One failure mechanism that is not typically designed for can occur with heavy excavation equipment loads passing over a low-rise arch with little soil cover. It is a single-shear failure mode on a transverse section through the arch, wall, and footing, in a radial pattern on each side of the wheels. Such a mode can occur before bending or punching shear failure with very thin-shell arch culverts.

21.9 Load Tests

Confirmation of the adequacy of many arch culvert and bridge designs has been obtained by full-scale load tests conducted to satisfy the following load test criteria: Service LL Deflections < Span/800 for Service Axle Load = 32 k (1 + IM) where IM is the dynamic load factor 0.33(1-fill height/8) (AASHTO, 2.5.2.6.2). A performance limit could be added by stating that cracks are to be less than 0.01 in wide.

Designs based on finite element soil–structure interaction analyses are verified by these load tests (Beach, 1988; Simpson Gumpertz & Heger, Inc., 2001). A compilation of field tests conducted on many arch culvert products is shown in Table 21.1. The load test results show that such a monolithic concrete system has a very high degree of redundancy and will provide the reserve strength (factor of safety) required to support severe overloads. In general, total elastic rebound occurred for the applied loads on the load–deformation curve.

21.10 Design Comments

21.10.1 Box Culverts

A box culvert works by providing serviceability due to low deflections and is one of the best sections in being easy to employ by just noting the applicable state and federal standards for the size desired. Few failures have been encountered with such structures. Certainly they may be overdesigned and, since an arc can be circumscribed within and fully contained by a thick-wall box section, they may be subject to arch action. Thus the path of any load placed on the surface of the box takes the form of a strut from the load to the two sides, which, due to their soil support at these locations, act as reaction points. Thus the sections behave more with the "locking" behavior characteristic of arc sections, which, as they try to deflect, "lock up" in compression and yield sections in compression as well as bending and shear. For this reason, box sections are very proficient, though not efficient, in carrying load with minimal deflections. An arch culvert structure designed using the principles of soil–structure interaction is more efficient than a box designed as a rigid frame since it uses the soil both to carry a portion of the loads and to provide confinement support to the structure (Katona and Vittes, 1980).

21.10.2 Footings and Slabs

The slabs existing between the footings of culverts are generally thinner than the footings. Due to their thinness, less load in moment is transferred to this center section. With shallow fills, the DL is small and the LL is distributed along the section, there is a small bearing stress taken by the footing, and little deflection develops in the center slab section which acts more like a thin-grade control structure to the water flow through the structure.

On a deeply buried culvert the section acts more as an inclusion in an elastic half-space with the vertical force acting on the top of the arch from the traffic, and dead loads are more or less equal to the reaction force against the bottom of the footing–slab structure.

Table 21.1 Field load test results.

Construct co-product Project	Project number or (date)	# Brls	Span S (ft)	Ht (ft)	Thick t (in)	S/t	Cover (equipment) (ft)	Load	Axle load (kips)	Measured deflection (in)	Allow deflection (span/800)	Page # FS
Load test results – arch culvert structures												
Reinforced concrete box culvert												
4th Av 97-56	(5/00)	5	12	7	10	14	1	Concrete truck	64	0.010	0.18	28
Granite construction – aero-form circular shotcrete arch												
Camino de la Canoa	89-111	1	32	–	9	43	7.5	CAT 988B Loader	60	0.012	0.48	73
Camino de la Canoa	89-111	4	24	–	8	36	2	CAT 988B Loader	60	0.008	0.36	68
Con/span precast concrete arch (6 ft Sect length)												
Ohio	(10/92)	1	36	9	10	43	0	1–200 kip Jack	41	0.100	0.54	5
Phoenix Gunite – hydro-arch shotcrete												
Dove Mtn	(8/02)	5	11	4.5	6.5	20	5	CAT 623E Scraper	66	0.007	0.17	43
Dove Mtn	(8/02)	1	42	12	12	42	2	CAT 623E Scraper	66	0.048	0.63	22
Sun City, NV	(11/92)	2	22	7.25	8–12	26	2.5	CAT 631D Scraper	91	0.053	0.33	14
Dove Mtn	(8/02)	1	12	8	6.5	22	1	CAT 623E Scraper	66	0.058	0.18	5
Dove Mtn	(8/02)	1	12	4	6.5	22	2	CAT 623E Scraper	66	0.149	0.18	2
C L Ridgeway – con-arch shotcrete												
Tapo Canyon, CA	99-34	2	20	16	8	30	5	Concrete Truck	44	0.005	0.30	73
Rancho Cancion	96162	1	16	–	7	27	5	CAT 615C Scraper	48	0.008	0.24	40
Lowe Reserve, CA	98-14	1	42	–	12	42	4	CAT 970F Loader	49	0.027	0.63	31
4th Av	97-56	1	8	–	6	16	0	Concrete Truck	72	0.007	0.12	29
ASARCO, Ray Mine	96-27	1	47	–	11	51	17	HaulPak 830E Trk	555	0.030	0.71	24
Anthem, Phoenix	99-18	1	80	–	12	80	1	CAT 621F Scraper	68	0.089	1.20	22

(continued overleaf)

Table 21.1 (*continued*)

Load test results – arch culvert structures

Construct co-product Project	Project number or (date)	# Brls	Span S (ft)	Ht (ft)	Thick t (in)	S/t	Cover (equipment) (ft)	Load	Axle load (kips)	Measured deflection (in)	Allow deflection (span/800)	Page # 1 FS
Gowan I, NV	98-17	2	20	–	8	30	1	CAT 623E Scraper	66	0.022	0.30	22
Lowe Reserve, CA	98-14	1	24	–	6.5	44	7	Deere 624G Loader	20	0.010	0.36	22
Red Hawk (Cir)	96146	2	24	–	6	48	3	Concrete Truck	72	0.035	0.36	19
Benson Airport	99-13	1	16	–	6	32	4	CAT 623F Scraper	72	0.025	0.24	19
Rancho Vistoso	95-46	2	20	8	7-9	30	4	CAT 623 Scraper	67	0.034	0.30	16
Benson Airport	99-13	1	16	–	6	32	3	CAT 631F Scraper	96	0.040	0.24	15
Lowe Reserve, CA	98-14	1	42	–	12	42	2	CAT 970F Loader	49	0.057	0.63	14
Benson Airport	99-13	2	12	–	7	21	4	CAT 623F Scraper	70	0.025	0.18	14
Benson Airport	99-13	3	12	–	7	21	1.5	CAT 623F Scraper	70	0.030	0.18	10
Benson Airport	99-13	3	12	–	7	21	2	CAT 623F Scraper	72	0.033	0.18	10
S. 12th Avenue	99-25	1	8	–	6	16	1	Concrete Truck	64	0.020	0.12	9
Benson Airport	99-13	1	8	–	5	19	2	CAT 623F Scraper	72	0.024	0.12	9
Rancho Cancion	96162	2	12	3.5	6	24	2	CAT 613C Scraper	30	0.017	0.18	8
Daimler-Chrysler, Phoenix	99054	6	11	–	6	22	0.2	Concrete Truck	45	0.026	0.17	7
Benson Airport	99-13	1	12	–	5	29	2	CAT 623F Scraper	72	0.050	0.18	6
Pinnacle Reserve II	96-28	4	28	–	9	37	1	CAT 966F Loader	27	0.044	0.42	6
Rancho Vistoso	95-46	2	16	4	7-9	24	3	CAT 966E Loader	27	0.032	0.24	5
HITEC, Conc Str	(11/00)	1	28	6.5	9	37	1	2–200 kip Jacks	187	0.420	0.42	5

HITEC, Shot Str	(11/00)	1	28	6.5	9	37	1	2–200 kip Jacks	159	0.420	0.42	4
Silverbell Hills	97157	3	12	4	6	24	1.5	CAT 966E Loader	49	0.057	0.18	4
Silverbell Hills	97157	3	12	–	6	24	1.5	CAT 627B Scraper	56	0.067	0.18	4
Thomp Pk Pkwy, Phoenix	99-12	2	28	–	9	37	2	10 000g Water Trk	56	0.160	0.42	4
Benson Airport	99-13	1	12	–	5	29	2	CAT 623F Scraper	79	0.130	0.18	3
Cable concrete – Integra-Arch concrete												
Verde Ranch	(8/04)	1	24	5.5	8	36	0.5	CAT 613C Scraper	26	0.015	0.36	15
Verde Ranch	(8/04)	1	34	6.5	8	51	1.5	CAT 613C Scraper	26	0.026	0.51	13
Shops at Thomp Pk	(10/02)	2	12	3.75	8	18	0.5	CAT 613C Scraper	30	0.013	0.18	10
RanchoParaiso-**StrOnly**	(9/04)	1	24	5	8.5	34	0	CAT 615C Scraper	45	0.070	0.36	5

Allowable deflection = span/800 defined for AASHTO 32 k **Service** axle LL (AASHTO) (2.5.2.6.2)

Service axle LL = LL*η*(1 + IM)*m

Impact factor, IM = 0.33(1-Fill Ht/8) (3.6.2.2)

Presence factor, m = 1.0 If no overlapping of wheel or axle load distribution (3.6.1.1.2)

Load modifier, η = 1.0 For service load (1.3)

FS (on deflection) = (axle Ld/meas deflt)/(32k × (1 + IM) × m × eta/(span/800))

FS for mine traffic loadings defined for mine truck design axle LL

The arched top section provides a good configuration for distributing this load to the walls by compression, resulting in less moment and minimal shear compared to a flat slab section. The base resultant pressure distribution will depend on the stiffness ratios of the slab and footing structures and the soil foundation. Since it is not arched, the bottom slab will crack and deflect upward as the footing settles unless it is well reinforced. Large shear can develop at the footing–wall juncture.

Thickening the slab portion of the footing to develop resistance is costly. It may be advantageous to cut the slab loose from the footing and let it float free or eliminate it entirely. The optimal design for the bottom section would be an arched shape that would be an arched configuration which yields compression in the section less bending and deflections.

21.10.3 Wall Sections

In a manner similar to the shear that develops at the footing–wall interface, the shear also develops at the wall–arch interface juncture. This shear in the arch–wall juncture can be reduced if the soil immediately adjacent to the top slab portion of the structure is either strong enough or strengthened to carry most of the thrust. The flatter the arch height, the greater the lateral thrust that is developed, causing less moment to be developed in the structure. This increased thrust for flatter slopes creates more confining pressure on the structure. The beneficial aspects of thrust must be optimized against the increased soil bearing stress and the increased shear in the wall when designing an arch-shaped form.

An arch section with thickened haunch section near the wall combines the benefits of an arch and a cantilever-type bridge. The thickened wall sections which are needed for tall walls and deep fill situations could also be more optimally designed by arching the thin-slab section on the exterior soil side.

The optimal design for a culvert becomes an arched top and bottom section with side portions that are less arched since the lateral pressure on the sides is approximately half of the vertical pressure except at the soil reaction point where it may reach a factor of 2 times the vertical soil pressure. Such a design when optimized to the fullest becomes a circular pipe shape.

21.11 Buckling of Arches

Buckling only occurs with buried arches that are characterized as flexible. Flexible arches have the potential to fail by a local buckling mechanism while concrete arches that are generally classified as rigid can be expected to fail in shear by a combination of bending, shear, and compression. The experimentally derived diameter, D, to thickness, t, ratios that tend to control the mode of failure are:

Rigid (no buckling):	$D/t < 100$
Flexible:	$D/t > 100$

Typical reinforced concrete buried barrel arches used for culverts or small bridges will lie well within the rigid criteria with a D/t of perhaps 40, while the comparable metal arch D/t will lie in the range of 125 (where t is considered as the effective metal thickness of the corrugations). Flexible structures for which $M_s/(EI/D^3)_{\text{structure}} > 10^4$ may be prone to buckle. For reinforced concrete section configurations, local buckling instability of thin-arch sections is not possible due to their geometrical section properties.

Both Timoshenko and Gere (1961, Section 7.6 Buckling of a Uniformly Compressed Circular Arch, pp. 297–302) and Allgood (1972, Section 4.4.1.6 Buckling, pp. 69–70) addressed buckling of cylinders, shells, and fixed-end arches. Timoshenko and Gere's equations are in conformance with the results that Allgood published but their solutions require tabulated buckling parameter values.

21.12 Seismic Design Considerations

Seismic forces from earthquakes are normally not considered in soil-structure interaction systems. Underground structures must move with the surrounding soil during earthquakes and usually will be supported by the interacting earth against crushing or collapse even if the structure's joints are strained.

(CALTRANS, 1994, BDP, p. 6-2)

In essence, even shallow-buried arch structures are buried structures, because the arch structure is "buried" by the soil comprising the spandrel fill zone above the arch and below the roadway. The deeper a structure is embedded in the ground and the stronger the surrounding soil material the less damage a structure will suffer. The seismic ground motions that are one to four times the tunnel diameter will be magnified when they act on a tunnel (Hashash *et al*, 2001).

Unlike surface structures which are designed for the inertial forces caused by ground accelerations, underground structures are designed for the ground deformations and strains that are imposed on the structure. A simple approach ignores the interactions between the structure and the enveloping ground and designs the structure for the free-field ground deformations. In more sophisticated approaches, such as a pseudo-static or dynamic analysis, the ground deformations are imposed as loads on the structure.

Often it is necessary to determine the period of a structure, T, from an acceleration–response–spectra (A-R-S) chart with peak rock acceleration. The weight of the structure and soil fill in the prism above the arch structure, W, in units of kips, can be determined. The stiffness of the soil–structure system, K, in units of kips/in, comprises the structure and the adjacent soil and the lateral soil resistance in the earth above the crown of the arch structure. The stiffness of the structure and its adjacent soil can readily be ascertained using a simple beam–bar model with a unit lateral load applied to its crown of the structure. The lateral stiffness of the soil above the structure would be equal to the coefficient of horizontal subgrade reaction times the height of the fill zone above the crown of the structure.

Using the specified units, the period of the soil–structure system, T, in seconds is

$$T = 0.32(W/K)^{0.5}$$

The natural frequency of the soil–structure system, f, in hertz is

$$f = 1/T$$

Entering into the A-R-S chart the maximum peak ground (rock) acceleration, a_{max}, the depth of alluvium, and the soil–structure system period, T, yields the modified A-R-S ground motion acceleration. For buried arch structures, their periods are less than 0.2 seconds, yielding natural frequencies above 5 Hz. This dynamic condition results in little amplification of the ground input acceleration.

 Related Workbooks on DVD

GeomShapes
T-CircEllipseCat&Par
T&MparLeontovichEqns

References

AASHTO (2010) *Bridge Design Specifications*, 5th edn, American Association of State Highway and Transportation Officials.

Allgood, J.R. (1972) Summary of Soil- Structure Interaction. Technical Report R 771, Naval Civil Engineering Laboratory, Port Hueneme, CA, July, Section 4.4.1.6 Buckling, pp. 69–70.

Bakht, B. (1997) Full use of arching in deck slabs. *Civil Engineering*, **57** (6), 14A–16A.

Beach, T.J. (1988) Load Test Report and Evaluation of Precast Concrete Arch Culvert (Con/Span), Preprint Paper No. 870205, Transportation Research Board, Washington, DC.

Beer, F.P. and Johnston, E.R. (1962) *Statics and Dynamics*, McGraw-Hill, Specifically Sections 7.8–7.9, Parabolic and Catenary Cables, pp. 258–268.

Caltrans (1994) *Bridge Design Practice Manual*, California Department of Transportation, p. 6–2.

Csagoly, P.F. and Lybas, J.M. (1989) Advanced design method for concrete bridge deck slabs. *Concrete International*, **11** (5), 53–63.

Hashash, Y.M.A. *et al.*, (2001) Seismic design and analysis of underground structures, *Tunnelling and Underground Space Technology*, Pergamon, vol 16, pp. 247–293.

Heyman, J. (1982) *The Masonry Arch*, Wiley-VCH Verlag GmbH, Specifically the catenary arc described by the Inglis Equation, pp. 46–49.

Katona, M.G. and Vittes, P.D. (1980) Soil-structure analysis and evaluation of buried box-culvert designs. *Transportation Research Record*, (878), 1–7.

Leontovich, V. (1959) *Frames and Arches: Condensed Solutions for Structural Analysis*, McGraw-Hill.

Simpson Gumpertz & Heger, Inc. (2001) *Summary of the Structural Evaluation of the Con-Arch System*, Highway Innovative Technology Evaluation Center (HITEC, CERF, ASCE), Arlington, TX.

Timoshenko, S.P. and Gere, J.M. (1961) *Theory of Elastic Stability*, McGraw-Hill, Section 7.6 Buckling of a Uniformly Compressed Circular Arch, pp. 297–302.

Van Beek, G.W. (1987) Arches and vaults in the ancient Near East, *Scientific American*, **257** (1), 46–53.

Part Five
Engineering Applications

"The most powerful force in the universe is compound interest."

– Albert Einstein

Part Five
Engineering
Application

22

Domes

22.1 Geometry

A dome surface is created by revolving an arch around its vertical axis. The geometry of a dome is developed by using either a circular or parabolic shaped arch. The top of the dome can be left open creating a circular-shaped open-topped or truncated dome (Corkhill *et al.*, 1974; Petroski, 2011). A dome is a very stable structure and is able to resist loads applied in any direction. It is like an arch supported in the out-of-plane direction.

22.2 Membrane Stresses

The stresses in a dome are due to compressive arch action in the radial (vertical) direction. Additionally, circumferential (denoted as tangential) stresses exist. These stresses are shown in Figure 22.1. Bending moments can be ignored for large radius/ height ratios.

The vertical stresses in a hemispherical-shaped dome will resolve to a thrust that is vertical. Domes less than a hemisphere can develop large outward thrusts at their base. This thrust can be resisted by a tension ring beam at their base. Historical structures used massive buttresses to resist this force.

In the top portion of a dome the circumferential forces are in compression and in the bottom portion they are in tension. As shown in Figure 22.2, a transition line inclined from $45°$ to $60°$ to the vertical denotes the separation of these forces. For a uniform dead load this transition line is inclined at $51.83°$ to the vertical. If there is an opening

Solutions for Soil and Structural Systems using Excel and VBA Programs, First Edition. Robert L. Sogge.
© 2012 John Wiley & Sons, Ltd. Published 2012 by John Wiley & Sons, Ltd.

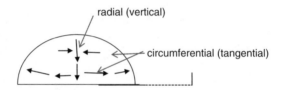

Figure 22.1   Membrane stresses in dome structure.

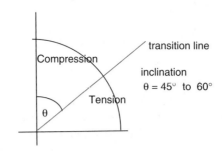

Figure 22.2   Tension–compression zones.

at the top of the dome the circumferential compressive forces need to be resisted by a compression ring. The most efficient way to resist the tension force caused by the circumferential compressive forces at the base of the dome is by using the same tension ring beam that resists the outward component of the vertical thrust. The support forces at the bottom of a dome consist of the footing for the vertical forces and a tension ring to resist the outward thrust and circumferential tension forces.

 ## 22.3 Stress Computations Using Worksheet Dome

The stresses in full or truncated domes that are formed by circular radii arcs, either with or without a circular opening at the top, are computed in the **Dome** worksheet. The N_ϕ, vertical (radial), and N_θ, tangential (circumferential), membrane forces at any point in the dome due to (i) self-weight dead load, (ii) a circular line load, force/ft, around the top of the dome, and (iii) a uniform internal pressure are determined (Timoshenko and Woinowsky-Krieger, 1959). Reinforcing weights and concrete quantities in the shell, slab, and footing are computed by the **Dome** worksheet.

The uniform internal pressure loadings can simulate the stress of inflatable forms that are used to create a surface from which the reinforcing steel can be hung and shotcrete blown against. This loading creates a tension state in the fabric "balloon."

Related Workbook on DVD

Dome

References

Corkill, P.A., Puderbaugh, H., and Sawyers, H.K. (1974) Domes, in *Structure and Architectural Design*, 2nd edn, Sernoll, pp. 151–164.

Petroski, H. (2011) Arches and domes. *American Scientist*, **99** (2), 111–115.

Timoshenko, S. and Woinowsky-Krieger, S. (1959) Particular cases of shells in the form of surfaces of revolution and Approximate methods of analyzing stresses in spherical shells, in *Theory of Plates and Shells*, 2nd edn, McGraw-Hill, pp. 436–437 and 547–554.

23

Critical Path Method

23.1 Project Scheduling

The critical path method (CPM) of scheduling provides the interdependency or inter-relationship of the tasks (activities) required to complete a project (Pinnell, 1980).

Project scheduling by the CPM is performed by the worksheet CPM that contains a VBA macro Sub Procedure labeled **CPM**. The VBA program is based on a BASIC program by G. Whitehouse presented in *Civil Engineering* in May 1981 (Whitehouse, 1981). This worksheet uses input data from cells in the worksheet. Output results are sent to the same worksheet. Additionally the macro creates a data file of the input and output data.

This program, although simplistic, does teach the fundamentals of the critical path scheduling method. For more complex projects, Microsoft Project, or the more advanced Primavera P6, may be required.

Program input, consisting of a schedule of the tasks required to complete a job, must fit within the time frame for the project. The tasks must be laid out in their order of sequence of initiation and completion. Some tasks may begin and/or end simultaneously. The duration of an activity must be estimated. It must include the contingency time associated with that specific task.

Program input

Title = Project description
N = Total number of work Activities or Tasks
J = Activity number
F = Node defining start of Activity
T = Node defining end of Activity
C = Duration of Activity
ACT$ = Description of Activity

Solutions for Soil and Structural Systems using Excel and VBA Programs, First Edition. Robert L. Sogge.
© 2012 John Wiley & Sons, Ltd. Published 2012 by John Wiley & Sons, Ltd.

Program output:
ES = Earliest Start
LS = Latest Start
TF = Total Float
FF = Free Float
Critical Path −indicated by**
Activity Description
Project Duration = total length of all activities on the critical path.

Float time for an activity is the time after a task is completed and before the next task begins. Free float time for an activity is the time one activity can be delayed before it impacts on the early starting time of the next activity. Total float time for an activity is the amount of slack in the start or completion time of an activity without delaying the project. Those activities that have zero float time are on the "critical" path.

The "critical path" is formed from the sequence of tasks that determine the project duration. Any delay in a task on this path directly effects the project duration and thus is "critical." The computed time for project completion must be compared to the total time allotted for its completion. Project completion time can be shortened by determining those tasks that can be performed simultaneously.

23.2 VBA Versions

There are two versions of this workbook saved as **CPMa** and **CPMb** both having the same **CPM** macro. The only difference between the two versions is how the macro is stored. In the **CPMa** worksheet the macro is stored as an Excel object in the location Sheet1.CPM (Sheet 1 is given the name of the Excel worksheet file CPM) as viewed by Project Explorer and not in a module. In the workbook **CPMb** the VBA macro application Sub CPM() is stored in a module location and, similar to the other version, it is executed using the Start button in the worksheet that is associated with the macro sub named CPM. Note that the macro consists of only one Sub Procedure that is contained as an object in the workbook.

References

Pinnell, S.S. (1980) Critical path scheduling: an overview and a practical alternative. *Civil Engineering*, July, 66–70.
Whitehouse, G.E. (1981) Critical path program for a micro-computer. *Civil Engineering*, May, 54–56.

24

Financial Analysis

24.1 Equations Governing Financial Operations

The equations for computing compound interest, sinking funds, and loans (funded annuity), along with their respective inverses of present value (without payment), future value (annuity), and present value (with payment), are as follows (Merriman and Wiggin, 1952):

Formula	Equation number
Amount of $1 at compound interest	
$a = p(1+i)^n$	(I)
Principal present value of $1 due at a future date	
$p = a/(1+i)^n$	$(IV) = 1/(I)$
Payment required to sinking or amortization fund to fund $1 annuity	
$\text{pmt} = ai/\{(1+i)^n - 1\}$	$(III) = 1/(II)$
Amount of annuity funded by $1 payments (future value of annuity)	
$a = (\text{pmt})\{(1+i)^n - 1\}/i$	(II)
Payment required to fund annuity of $1 value (loan)	
$\text{pmt} = pi/\{(1 - (1+i)^n\}$	$(VI) = 1/(V) = (I)^*(III) = (I)/(II)$
Principal present value of annuity paying $1	
$p = (\text{pmt})\{1 - (1+i)^{-n}\}/i$	$(V) = 1/(VI) = (II)^*(IV) = (II)/(I)$

where a = amount, p = principal, i = interest, n = term, and pmt = payment.

These equations have been derived for end-of-term payments.

Solutions for Soil and Structural Systems using Excel and VBA Programs, First Edition. Robert L. Sogge.
© 2012 John Wiley & Sons, Ltd. Published 2012 by John Wiley & Sons, Ltd.

24.2 Excel Worksheets for Financial Calculator and Formulas

The financial calculator provided in Worksheet **LoanEqns** of **ProgLoan** solves these equations for the variables of amount, principal, interest, term, and payment. Given any three of these variables, the fourth is calculated. The way the spreadsheet is written, the unknown variable, once computed, can then be entered and all variables checked for conformance with the input values.

A calculation for the remaining principal balance of any loan after a specific term length is made using two different equations. Another calculation is made, using the amortized loan payment amount, to compute the number of payments required to pay off the principal and then those required to pay off the accumulated interest. Such a calculation is handy when structuring a loan for which the principal is paid first, and if and when it is repaid, the interest follows.

For the loan equation spreadshcet, the interest on the loan, if unknown, is calculated from an iterative trial and error procedure written as a macro Function in VBA. In this macro the interest is incremented by 0.0001 until closure. Values of variables in the spreadsheet are passed to the macro Function in the enclosed parentheses, and the value of the yearly interest is returned by the Function name, YRINT. For the sinking fund equation spreadsheet, the yearly interest on the loan, if unknown, is calculated from an iterative procedure written in a VBA macro Function SEINT.

Financial amounts can be computed using the Excel-provided Financial Functions PV() present value, FV() future value, RATE() interest rate, PMT() amount of payment, and NPER() number of payment periods. These functions are employed in the worksheet labeled **LoanEqns** on the tab. When using Financial Functions to compute compound interest beginning and ending amounts, use the present and future value functions of these amounts with payments equal to zero.

A worksheet for loan payment schedules (**LoanPmtSchd**), in which the payment per period, the term, or daily interest is computed, is presented and a worksheet for a loan payoff amount (**LoanPayoffAmt**) for various values of interest payments per year is provided. The **Formulas** worksheet provides a worksheet for calculating values of financial equations. The worksheet tabs **LoanPmtSchd, LoanSchd, DailyLoanSchd,** and **LoanPayoffAmt** provide amortization schedules and other financial calculations.

24.3 Significant Aspects of Excel Worksheet and Macro Functions

In order to have only the one unknown appear in cells F8–F11 of the spreadsheet, a statement similar to the following should be inserted in those cells:

```
=IF(D8=" ",D11*(1-(1+D9/D12) ^ (-D10*D12))/(D9/D12)," ")
```

As can be seen from the equations supplied in the worksheets, when giving a variable a percentage format it is not necessary to divide the number by 100 to turn it into a decimal. In other words, 20% of 10 equals 2.

The workbook **LoanProg** has a good example of a Function Procedure that calculates interest by an iterative routine. An example of a function would be the following:

```
Function C(As Single, B As Single), As Single
C = SQRT(A+B)
End Function
```

The statement Function C() in a cell of the worksheet returns the value to that cell. This approach is different from a Sub Procedure that does not return a value.

 Related Workbook on DVD

LoanProg

Reference

Merriman, T. and Wiggin, T.H. (1952) *American Civil Engineers' Handbook*, 5th edn, John Wiley & Sons, Inc., Fourteenth Printing (Specifically Section 2, Article 7, Interest and Sinking Fund, pp. 55–58).

25

Conversion of Units
of Measurement

25.1 Unit Systems

The units used in this text are US Customary Standard (USCS) since conversion to the International System of Units (SI), a modern metric system, never really made it in the United States. Conversion of the various units of measurement and their dimensions can be very confusing when the USCS system and (fps) or ft–lb–sec are involved.

All unit systems are based on defined quantities for length, mass, time, and temperature. Defined units are denoted as basic units and consist of the following for all units systems:

Length (L)
Mass (M)
Time (T)
Temp (temp).

The following units can be derived from the basic units and are known as derived units:

$$Area = Length^2 = L^2$$
$$Volume = Length^3 = L^3$$
$$Moment\ of\ Inertia = L^4/L$$
$$Acceleration = L/T^2$$

Solutions for Soil and Structural Systems using Excel and VBA Programs, First Edition. Robert L. Sogge.
© 2012 John Wiley & Sons, Ltd. Published 2012 by John Wiley & Sons, Ltd.

Force $= F =$ Mass $\times$ Acceleration $= ML/T^2$

Pressure (Stress) $=$ Force/Area $= F/L^2 = (ML/T^2)/L^2 = M/(T^2L)$

Moment or Energy $=$ Force $\times$ Length $= FL = ML^2/T^2$

Power $=$ Energy/Time $= FL/T ML^2/T^3$

Unit Weight $=$ Force/Volume $= F/L^3$

Velocity (Permeability) $= (L/T)$

Flow Rate $= (L^3/T)$.

25.2 Defined Units

The basis for the USCS system is the defined units, expressed in the units shown, of:

in $= 25.4$ mm

$lb_m = 0.453\ 592\ 37\ kg_m$

and the useful defined quantity of gal $= 231$ in^3

from which all other units in the USCS can be derived.

The defined units in the SI are expressed in the units shown of:

Gravitational acceleration $= g = 9.806\ 65\ m/s^2$

Atmospheric pressure $=$ atm $= 101.325\ kPa = 760\ mm\ Hg$

from which all other units in the SI can be derived.

25.3 Labeling Conventions

In the SI, the units start with lower case letters, except when they are named after a person. For example, the newton is denoted as N. The only exception is the symbol for liter which is a capital L (but not always).

In the SI all units start with uncapitalized letters with the following two exceptions:

1. When the symbol is an abbreviation for a person's name, that is, Celsius (C), Joule (J), Newton (N), Pascal (Pa), Watt (W).
2. The liter is L not l.

The case of the prefix symbols, indicating exponential multipliers, changes depending on their value:

Prefix	Abbreviation	Exponent 10^0
yotta	Y	24
zetta	Z	21
exa	E	18
peta	P	15
tera	T	12
giga	G	9
mega	M	6
kilo	k	3
hecto	h	2
deca	da	1
deci	d	−1
centi	c	−2
milli	m	−3
micro	μ	−6
nano	n	−9
pico	p	−12
femto	f	−15
atto	a	−18
zepto	z	−21
yocto	y	−24

The hectare (symbol ha) adds some confusion to the prefix system. It is a unit of area equal to $10\,000\,m^2$ ($107\,639.1\,sq\,ft$), or $1\,hm^2$ ($0.1\,km$, or $100\,m$, squared). The symbol for hecto or hecta is h and the prefix denotes the factor 10^2 or 100. This dimension is commonly used in the SI by the surveying profession for measuring land area. Its base unit, the area, was defined by older forms of the metric system. It is no longer part of the modern metric system.

 ## 25.4 Workbook UnitCnvrsn

One should always attach units to Excel workbooks and VBA procedures as they are often ignorant of units and problems could arise. Usually the best manner of VBA programming is to make units input = units output.

The workbook **UnitCnvrsn** converts units between the USCS and SI systems. It is unique in that all derived units are computed from the basic definitions, which may be in the USCS or in SI. Approximate conversion factors are also presented in this worksheet.

25.5 Excel Conversions

Within Excel there is a function that converts a number from one measurement system to another. The function label is

```
= convert(number,"from unit","to unit"), for example
= convert(1.0,"lbm","kg").
```

This function is entered using the insert Function button under the Formulas tab. This function is available if the analysis ToolPak add-in is installed. Nested Convert functions are needed for square length and other units. In Excel 2003 there exist some minor conversion discrepancies.

25.6 Example

25.6.1 Example: ksf (1000 lb/ff²) to Pascals

The ksf units are not included in the Excel Convert function; the only pressure units are the Pascal, atmosphere, and mm of Hg. Thus the worksheet becomes very useful for the many conversions not handled directly by the Excel Convert function.

 Related Workbook on DVD

UnitCnvrsn

Index

Solutions for Soil and Structural Systems using Excel and VBA Programs, First Edition. Robert L. Sogge.
© 2012 John Wiley & Sons, Ltd. Published 2012 by John Wiley & Sons, Ltd.